Probability Models *in* Operations Research

The Operations Research Series

Series Editor: A. Ravi Ravindran
Dept. of Industrial & Manufacturing Engineering
The Pennsylvania State University, USA

Integer Programming: Theory and Practice
John K. Karlof

Operations Research: A Practical Approach
Michael W. Carter and Camille C. Price

Operations Research Calculations Handbook
Dennis Blumenfeld

Operations Research and Management Science Handbook
A. Ravi Ravindran

Probability Models in Operations Research
C. Richard Cassady and Joel A. Nachlas

Forthcoming Titles

Applied Nonlinear Optimization Modeling
Janos D. Pinter

Operations Research Calculations Handbook, Second Edition
Dennis Blumenfeld

Probability Models *in* Operations Research

C. Richard Cassady

Joel A. Nachlas

CRC Press
Taylor & Francis Group
Boca Raton London New York

CRC Press is an imprint of the
Taylor & Francis Group, an **informa** business

Reprint 2025
CRC Press
Taylor & Francis Group
6000 Broken Sound Parkway NW, Suite 300
Boca Raton, FL 33487-2742

Library of Congress Cataloging-in-Publication Data

Cassady, C. Richard.
Probability models in operations research / C. Richard Cassady and Joel A. Nachlas.
p. cm. -- (Operations research series ; 5)
Includes bibliographical references and index.

1. Operations research--Mathematics. I. Nachlas, Joel A. II. Title. III. Series.

T57.6.C378 2008
658.4'034--dc22 2008023731

Visit the Taylor & Francis Web site at
http://www.taylorandfrancis.com

and the CRC Press Web site at
http://www.crcpress.com

ISBN 978-0-367-24244-2 (pbk)

Printed and bound in India

For sale in India, Pakistan, Nepal, Bhutan, Bangladesh and Sri Lanka only.

Contents

To Wendy Privette-Cassady and Beverley Nachlas,
two rocks of stability in a world of uncertainty

Preface

This book is intended for use as a textbook in one-semester courses that introduce upper-level undergraduate or entry-level graduate industrial engineering students to the concepts of probabilistic and stochastic models in operations research. In this text, we cover basic concepts of probability and distribution theory, conditional probability and expectation, renewal processes, Bernoulli processes, the Poisson process, discrete-time and continuous-time Markov chains, and the basic concepts of Markovian queueing systems.

Before taking this course, students should have an understanding of the basic principles of multivariable calculus, linear algebra, and differential equations. If undergraduate students have completed a traditional undergraduate course or courses in probability and statistics, then the instructor can focus more on conditional probability and stochastic processes. Graduate students who have completed a one-semester course in probability and distribution theory are the best prepared for and gain the most from this material.

Regardless of whether this text is used for undergraduate or graduate students, we believe that the topics in all eight chapters can be covered in one semester. In an undergraduate course, the instructor can limit focus to concepts and examples. In a graduate course, the instructor should have time to also focus on proofs and derivations. With this in mind, we leave most of the proofs and derivations for homework problems.

To the extent possible, we present examples and homework problems in an industrial engineering context. In doing so, we are true to the fact that industrial engineering has expanded from its origins in manufacturing and into transportation, health care, logistics, service, etc. To show how the concepts presented in this text come together in the analysis of meaningful problems, we conclude each of the first six chapters with an extended homework problem, which we refer to as an *Application*. In our own experience in teaching from the text, students have found these applications to be the most difficult and rewarding part of the course. The seventh and eighth chapters of this book do not have applications, because in our opinion, Chapter 8 is *the* appropriate application for the concepts from Chapter 7.

When we teach from this text, we use a subset of the homework problems for in-class examples and another subset for actual assigned homework problems. We are happy to provide these subsets to interested faculty. A *Solutions Manual* for this text is also available. To instructors who either adopt or consider adopting this text, we welcome your comments and suggestions for improvement.

Finally, we would like to thank those people who have facilitated the completion of this text: John Kobza (the original author of the Chapter 6 application), Ed Pohl, Ebru Bish, Pingjian Yu, Yisha Xiang, and many, many current and former students at the University of Arkansas and Virginia Tech.

C. Richard Cassady
Joel A. Nachlas

Authors

C. Richard Cassady is Associate Professor of Industrial Engineering and Director of Freshman Engineering at the University of Arkansas. He joined the Department of Industrial Engineering at the University of Arkansas in August 2000. During 2006 to 2007, he held the John L. Imhoff Chair in Industrial Engineering. In 2005, he was named Industrial Engineering Faculty Member of the Year by the Arkansas Academy of Industrial Engineering. Prior to joining the University of Arkansas faculty, he was an Assistant Professor of Industrial Engineering at Mississippi State University (1996–2000).

Dr. Cassady's primary research interests are in the application of probability, optimization, simulation, and statistics to problems related to evaluating, improving, or optimizing the performance of repairable equipment. This work includes the analysis and development of equipment maintenance policies. He also conducts research in the areas of reliability engineering, statistical quality control, logistics systems modeling, and sports applications of operations research. He has received several research awards, including the 1999, 2001, and 2006 Stan Ofthsun Awards from the Society of Reliability Engineers, and the 2004 Institute of Industrial Engineers (IIE) Annual Conference Quality Control and Reliability Best Paper Award.

Dr. Cassady's primary teaching interests are in reliability and maintainability engineering and probabilistic operations research. He also teaches courses in basic probability and statistics, statistical quality control, and deterministic operations research. He previously served as the instructor for the introductory Industrial Engineering course and the Industrial Engineering capstone design course. He is an elected member of the University of Arkansas Teaching Academy, and he has received numerous teaching awards, including the Charles and Nadine Baum Faculty Teaching Award from the University of Arkansas (2006), the inaugural Imhoff Outstanding Teacher Award from the University of Arkansas College of Engineering (2005), and the 2004 Alan O. Plait Award for Tutorial Excellence from the Annual Reliability and Maintainability Symposium.

Dr. Cassady is a Senior Member of IIE and a member of Tau Beta Pi, Alpha Pi Mu, the American Society for Engineering Education (ASEE), the Institute for Operations Research and the Management Sciences (INFORMS), and the Society of Reliability Engineers (SRE). He is a member of the Management Committee of the Annual Reliability and Maintainability Symposium, and he is an associate editor of the *Journal of Risk and Reliability*.

Dr. Cassady was born in and raised near Martinsville, Virginia. He received his B.S., M.S., and Ph.D. degrees, all in Industrial and Systems Engineering, from Virginia Tech. His wife, Wendy, is a speech therapist in private practice in Springdale, Arkansas, and they have two sons, Bryant (8/27/02) and Ramey (10/23/06).

Joel A. Nachlas serves on the faculty of the Grado Department of Industrial and Systems Engineering at Virginia Tech. He has taught at Virginia Tech since 1974 and acts as the coordinator for the department's graduate program in Operations Research. For the past twelve years, he has also taught Reliability Theory regularly at the Ecole Superiore d'Ingenieures de Nice-Sophia Antipolis.

Dr. Nachlas received his B.E.S. from the Johns Hopkins University in 1970, his M.S. in 1972, and his Ph.D. in 1976, both from the University of Pittsburgh. His research interests are in the applications of probability and statistics to problems in reliability and quality control. His work in microelectronics reliability has been performed in collaboration with and under the support of the IBM Corporation, INTELSAT, and the Bull Corporation. He is the co-author of over fifty refereed articles, has served in numerous editorial and referee capacities, and has lectured on reliability and maintenance topics throughout North America and Europe.

Dr. Nachlas's collaborative study with J. L. Stevenson of INTELSAT of the deterioration of microelectronic communications circuits earned them the 1991 P. K. McElroy Award. In addition, his work on the use of nested renewal functions to study opportunistic maintenance policies earned him the 2004 William A. Golomski Award.

Since 1986, Dr. Nachlas has served on the Management Committee for the Annual Reliability and Maintainability Symposium. Since 2002, his responsibility for the RAMS has been as Proceedings editor. He is a member of INFORMS, IIE, and SRE and is a Fellow of the American Society for Quality (ASQ).

During most of his tenure at Virginia Tech, Dr. Nachlas has also served as head coach of the men's lacrosse team and in 2001 he was selected by U.S. Lacrosse as the men's division intercollegiate associates national coach of the year.

1

Probability Modeling Fundamentals

The primary focus of industrial engineering is improvement. The origins of the field lie in the second stage of the industrial revolution with the move to mass production. The scope of application of industrial engineering has expanded to include banking and finance, health care delivery, military operations, retail goods distribution, communications networks, transportation systems, government services, and other avenues of human endeavor in addition to the traditional domain of manufacturing. Throughout this evolution in industrial engineering, the focus has remained process improvement, productivity, and effectiveness. As the discipline of industrial engineering has evolved, the set of tools that it encompasses has been refined and expanded. The methods of operations research have comprised an important ingredient in this expansion.

As in all engineering disciplines, an operations research study of an industrial system typically consists of constructing a mathematical or computer simulation model of a system, analyzing or experimenting with the model, and then based on the results, drawing conclusions about or making decisions regarding the operation of the system. Because the focus of operations research is on the development and analysis of these models, an operations research analyst must be aware of the key, inherent features of the system of interest that must be represented in the model. One such feature is **randomness**. All industrial systems are subject to various forms of randomness, and in many cases, this randomness is too significant to ignore. For example, the quality of a manufactured product, the time required to complete a surgical procedure, the lead time for a shipment of goods, and the customer demand for a certain product may all be subject to randomness.

But what is randomness? Common definitions of randomness feature concepts such as unpredictability, chance, and unrepeatability. So perhaps a more interesting question is: Why does randomness exist? One group of people would argue that randomness exists only due to our inability to observe or understand certain phenomena. These people are not necessarily predeterminists. They simply believe that everything (other than perhaps human choice) happens for a scientific reason. Another group of people would argue that even if you had the ability to observe and understand everything, you would still be occasionally surprised. Albert Einstein summarized this ongoing debate in a letter to Max Born: "You believe in a God who plays dice, and I in complete law and order."

Regardless of your opinion on the Einstein/Born debate, randomness is an important characteristic of many industrial systems that we would like to model. Just as geometry is the field of mathematics that we use to describe spatial relationships, probability is the component of mathematics that we use to model randomness. We use probabilistic and stochastic models to capture the random elements of a system. The fundamental models of this type are the subject of this book, and we begin our exploration of these models by presenting some essential concepts from probability theory.

1.1 Random Experiments and Events

The fundamental building block of probability theory is the random experiment.

> A **random experiment** is an occurrence such that the outcome cannot be predicted with certainty.

The experiment of interest may be a laboratory experimental trial, but it may also be a natural occurrence, a customer arrival, the manufacture of an item, etc. We complete the definition of a random experiment by delineating its sample space.

> The set of all possible outcomes of a random experiment is called the **sample space**. The sample space is denoted by Ω. An individual element of the sample space (some individual outcome) is denoted by ω.

Example 1.1

A manufactured product is subject to four potential flaws. The first three types of flaws can occur in any combination, but the fourth flaw can only occur in isolation. Consider the random experiment of producing a single item. Identify the sample space for this experiment.

$$\Omega = \{\text{none}, 1, 2, 3, 4, 12, 13, 23, 123\}$$

where none indicates that no flaws are present
- 1 indicates that only the first type of flaw is present
- 2 indicates that only the second type of flaw is present
- 3 indicates that only the third type of flaw is present
- 4 indicates that only the fourth type of flaw is present
- 12 indicates that only the first and second types of flaws are present
- 13 indicates that only the first and third types of flaws are present

23 indicates that only the second and third types of flaws are present
123 indicates that all of the first three types of flaws are present □

Example 1.2

When a customer orders an item, the customer must choose a model (basic or deluxe) and a color (maroon, orange, or white). Consider the random experiment of a single customer ordering one item. Identify the sample space for this experiment.

$$\Omega = \{BM, BO, BW, DM, DO, DW\}$$

where the first letter in each outcome indicates the model type (B implies basic, D implies deluxe) and the second letter in each outcome indicates the color (M denotes maroon, O denotes orange, W denotes white) □

In some cases, especially when the sample space is very large, the individual outcomes of the sample space provide more detail than the analyst needs. For these cases, we use events to combine the individual outcomes into meaningful groups.

> A subset of the sample space is called an **event**. Events are denoted by italicized, capital letters.

The particular events of interest depend on the system and problem under consideration. Therefore, clearly defining all the events of interest is a necessary step in building probability models.

Example 1.1 (continued)

Let F_i denote the event that flaw i is present in the manufactured product, i = 1, 2, 3, 4. Identify the individual outcomes that comprise F_1, F_2, F_3, and F_4.

$$F_1 = \{1, 12, 13, 123\}$$

$$F_2 = \{2, 12, 23, 123\}$$

$$F_3 = \{3, 13, 23, 123\}$$

$$F_4 = \{4\}$$ □

Example 1.2 (continued)

Let E_i denote the event that the customer orders color i, i = M (maroon), O (orange), W (white). Identify the individual outcomes that comprise E_M, E_O, and E_W.

$$E_M = \{BM, DM\}$$

$$E_O = \{BO, DO\}$$

$$E_W = \{BW, DW\} \square$$

Unless explicitly stated otherwise, we assume throughout this text that an event is non-empty. In other words, there is at least one outcome from the sample space in the event.

One graphical method of describing events defined on a sample space is the **Venn diagram**. The typical representation of an event using a Venn diagram is shown in Figure 1.1. In Figure 1.1, the rectangle corresponds to the sample space, and the shaded region corresponds to the event of interest.

Constructing probability models associated with a random experiment requires an understanding of several fundamental relationships between events. Discussion of these relationships begins with the definition of subsets and equivalence.

> If A and B are two events defined on a sample space Ω, then A is a **subset** of B, denoted by $A \subseteq B$, if and only if for all $\omega \in A$, $\omega \in B$. Note that A is a **proper subset** of B, denoted by $A \subset B$, if and only if $A \subseteq B$ and there exists some $\omega \in B$ such that $\omega \notin A$ (see Figure 1.2).

> If A and B are two events defined on a sample space Ω, then A is **equivalent** to B, denoted by $A = B$, if and only if $A \subseteq B$ and $B \subseteq A$.

For Example 1.1, we defined an event for the case in which the first type of flaw is present in the manufactured product. We may also be interested in the case in which this type of flaw is not present.

> If A is an event defined on a sample space Ω, then $\omega \in A^c$ if and only if $\omega \notin A$. The event A^c is called the **complement** of A (see Figure 1.3). Note that $\Omega^c = \phi$ (the **null** or **empty event**).

Example 1.1 (continued)

Identify the individual outcomes that comprise F_1^c, the event that the first type of flaw is not present in the product.

$$F_1^c = \{\text{none}, 2, 3, 4, 23\} \square$$

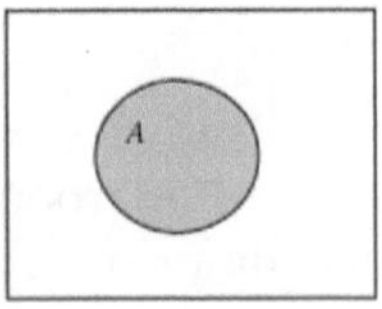

FIGURE 1.1
Venn diagram for event A.

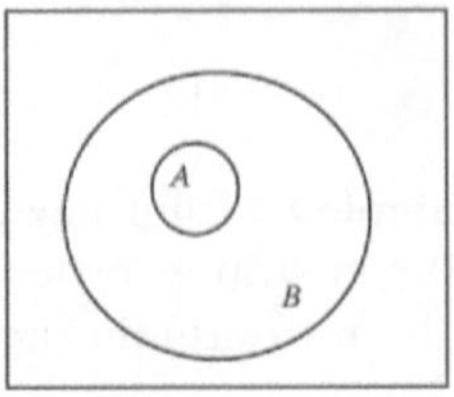

FIGURE 1.2
Venn diagram for $A \subset B$.

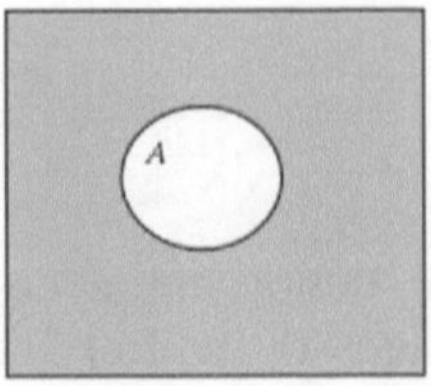

FIGURE 1.3
Venn diagram for A^c.

For Example 1.1, we may also be interested in the case in which any of multiple types of flaw is present.

> If A and B are two events defined on a sample space Ω, then $\omega \in A \cup B$ if and only if $\omega \in A$ or $\omega \in B$. The event $A \cup B$ is called the **union** of A and B (see Figure 1.4). If $\{A_1, A_2, \ldots\}$ is a collection of events defined on the sample space, then
>
> $$\omega \in \bigcup_{j=1}^{\infty} A_j$$
>
> if and only if there exists some $j \in \{1, 2, \ldots\}$ such that $\omega \in A_j$. The event
>
> $$\bigcup_{j=1}^{\infty} A_j$$
>
> is called the **union** of $\{A_1, A_2, \ldots\}$.

Example 1.1 (continued)

Let F denote the event that at least one flaw is present in the manufactured product. Express F in terms of the events F_1, F_2, F_3, and F_4, and identify the individual outcomes that comprise F.

$$F = F_1 \cup F_2 \cup F_3 \cup F_4 = \bigcup_{i=1}^{4} F_i$$

$$F = \{1, 2, 3, 4, 12, 13, 23, 123\} \square$$

For Example 1.1, we may also be interested in the case in which all of multiple types of flaws are present.

> If A and B are two events defined on a sample space Ω, then $\omega \in A \cap B$ if and only if $\omega \in A$ and $\omega \in B$. The event $A \cap B$ is called the **intersection** of A and B (see Figure 1.5). If $\{A_1, A_2, \ldots\}$ is a collection of events defined on the sample space, then

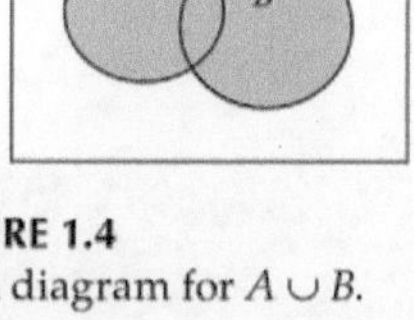

FIGURE 1.4
Venn diagram for $A \cup B$.

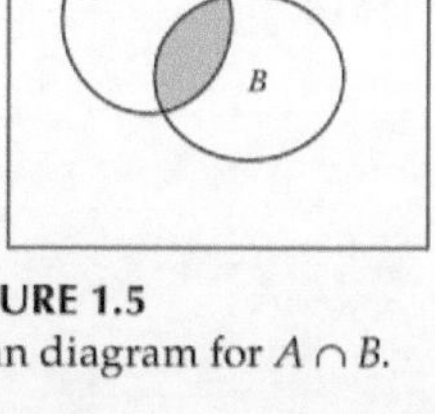

FIGURE 1.5
Venn diagram for $A \cap B$.

$$\omega \in \bigcap_{j=1}^{\infty} A_j$$

if and only if $\omega \in A_j$ for all $j \in \{1, 2, \ldots\}$. The event

$$\bigcap_{j=1}^{\infty} A_j$$

is called the **intersection** of $\{A_1, A_2, \ldots\}$.

Example 1.1 (continued)

Let E_{12} denote the event that the first and second types of flaws are present in the manufactured product. Express E_{12} in terms of the events F_1, F_2, F_3, and F_4, and identify the individual outcomes that comprise E_{12}.

$$E_{12} = F_1 \cap F_2$$

$$E_{12} = \{12, 123\} \square$$

The definitions of complements, unions, and intersections of events can be used to construct useful general relationships among events.

Let A, B, and C be events defined on a sample space Ω.

$$(A^c)^c = A \tag{1.1}$$

$$A \cup \phi = A \tag{1.2}$$

$$A \cup \Omega = \Omega \tag{1.3}$$

$$A \cup A^c = \Omega \tag{1.4}$$

$$\text{If } A \subseteq B, \text{ then } A \cup B = B \tag{1.5}$$

$$A \cap \phi = \phi \tag{1.6}$$

$$A \cap \Omega = A \tag{1.7}$$

$$A \cap A^c = \phi \tag{1.8}$$

$$\text{If } A \subseteq B, \text{ then } A \cap B = A \tag{1.9}$$

$$(A \cup B) \cup C = A \cup (B \cup C) \tag{1.10}$$

$$(A \cap B) \cap C = A \cap (B \cap C) \tag{1.11}$$

$$A \cap (B \cup C) = (A \cap B) \cup (A \cap C) \tag{1.12}$$

$$A \cup (B \cap C) = (A \cup B) \cap (A \cup C) \tag{1.13}$$

$$(A \cup B)^c = A^c \cap B^c \tag{1.14}$$

$$(A \cap B)^c = A^c \cup B^c \tag{1.15}$$

Equation (1.1) indicates that the complement of an event's complement is the original event. Equation (1.2) indicates that the union of an event and the empty event is the original event, whereas equation (1.6) indicates that the intersection of the event and the empty event is the empty event. Equation (1.3) indicates that the union of an event and the sample space is the sample space, whereas equation (1.7) indicates that the intersection of the event and the sample space is the original event. Equation (1.4) indicates that the union of an event and its complement is the sample space, whereas equation (1.8) indicates that the intersection of the event and its complement is the empty event. Statement (1.5) indicates that the union of an event and one of its subsets is the original event, whereas statement (1.9) indicates that the intersection of the event and one of its subsets is the subset. Equations (1.10) to (1.15) describe the distributive properties of unions, intersections, and complements.

Equation (1.8) establishes that an event and its complement have no common outcomes, that is, their intersection is empty. However, this lack of an intersection is not limited to complements. In Example 1.2, the events E_M and E_O have no common outcomes.

If A and B are two events defined on a sample space Ω, then A and B are said to be **mutually exclusive** or **disjoint** if and only if $A \cap B = \phi$ (see Figure 1.6). A collection of events $\{A_1, A_2, \ldots\}$ defined on the sample space is said to be **disjoint** if and only if every pair of events in the collection is mutually exclusive.

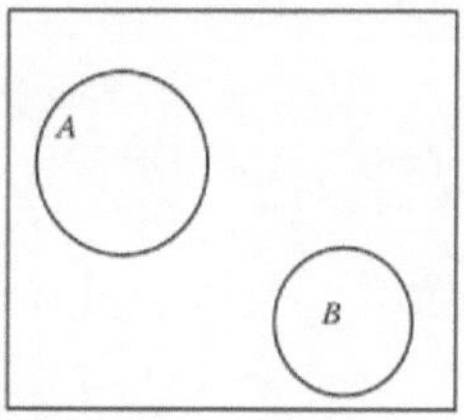

FIGURE 1.6
Venn diagram for mutually exclusive events.

Note that in Example 1.2, the collection of events $\{E_M, E_O, E_W\}$ is disjoint. Furthermore, note that the union of these events is the entire sample space.

A collection of events $\{A_1, A_2, \ldots\}$ defined on a sample space Ω is said to be a **partition** (see Figure 1.7) of Ω if and only if the collection is disjoint and

$$\bigcup_{j=1}^{\infty} A_j = \Omega$$

1.2 Probability

Probability is the measure used to quantify the likelihood of the events of a random experiment.

Let Ω be the sample space for some random experiment. For any event defined on Ω, Pr is a function which assigns a number to the event. Note that Pr is called the **probability** of the event provided the following conditions hold:

1. $0 \le \Pr(A) \le 1$, for any event A
2. $\Pr(\Omega) = 1$
3. If $\{A_1, A_2, \ldots\}$ is a disjoint collection of events, then

$$\Pr\left(\bigcup_{j=1}^{\infty} A_j\right) = \sum_{j=1}^{\infty} \Pr\left(A_j\right)$$

The three components of this definition, the **axioms of probability**, enable us to compute the probability of any event of interest. Also, these axioms can be used to derive several rules that might make the computations simpler.

Let A, B, C, and D be events defined on a sample space Ω.

$$\Pr(\phi) = 0 \tag{1.16}$$

$$\Pr(A) + \Pr(A^c) = 1 \tag{1.17}$$

$$\text{If } A \subseteq B, \text{ then } \Pr(A) \le \Pr(B) \tag{1.18}$$

$$\Pr(A \cup B) = \Pr(A) + \Pr(B) - \Pr(A \cap B) \tag{1.19}$$

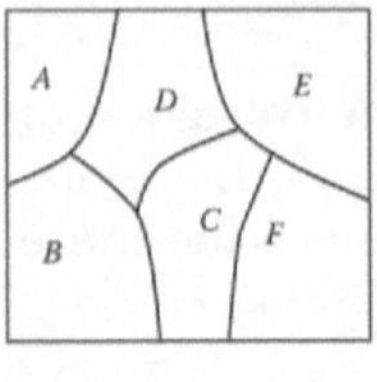

FIGURE 1.7
Venn diagram for a partition $\{A, B, C, D, E, F\}$.

$$\begin{aligned}\Pr(A \cup B \cup C) = \Pr(A) + \Pr(B) + \Pr(C) \\ - \Pr(A \cap B) - \Pr(A \cap C) - \Pr(B \cap C) \\ + \Pr(A \cap B \cap C)\end{aligned} \tag{1.20}$$

$$\begin{aligned}\Pr(A \cup B \cup C \cup D) = \Pr(A) + \Pr(B) + \Pr(C) + \Pr(D) \\ - \Pr(A \cap B) - \Pr(A \cap C) - \Pr(A \cap D) \\ - \Pr(B \cap C) - \Pr(B \cap D) - \Pr(C \cap D) \\ + \Pr(A \cap B \cap C) + \Pr(A \cap B \cap D) \\ + \Pr(A \cap C \cap D) + \Pr(B \cap C \cap D) \\ - \Pr(A \cap B \cap C \cap D)\end{aligned} \tag{1.21}$$

Equation (1.16) indicates that the probability of the empty event is zero. Equation (1.17) indicates that the sum of the probabilities for an event and its complement is one. Equation (1.18) indicates that the probability of an event must be at least as large as the probability of any one of its subsets.

Equation (1.19) extends the third axiom of probability to the case in which two events of interest are not disjoint. If the individual-event probabilities are summed, then the outcomes in the intersection are counted twice. Therefore, the probability of the intersection must be subtracted. This concept is extended to three and four events in equations (1.20) and (1.21), respectively, and it can be extended to even more events by observing the pattern in equations (1.19) to (1.21).

Example 1.1 (continued)

The probability that the first type of flaw is present in the manufactured product is 0.2. The probability that the second type of flaw is not present in the manufactured product is 0.8. The probability that the third type of flaw is present in the manufactured product is 0.17. The probability that the fourth type of flaw is present in the manufactured product is 0.01. The probability that at least one of the first two types of flaws is present in the manufactured product is 0.33. The probability that the second and third types of flaws are present in the manufactured product is 0.06. Compute the following probabilities.

(a) the probability that the second type of flaw is present in the manufactured product
From equation (1.17),

$$\Pr(F_2) = 1 - \Pr(F_2^c) = 1 - 0.8 = 0.2$$

(b) the probability that at least one of the second and third types of flaw is present in the manufactured product
From equation (1.19),

$$\Pr(F_2 \cup F_3) = \Pr(F_2) + \Pr(F_3) - \Pr(F_2 \cap F_3) = 0.2 + 0.17 - 0.06 = 0.31$$

(c) the probability that the first and second types of flaw are present in the manufactured product
From equation (1.19),

$$\Pr(F_1 \cap F_2) = \Pr(F_1) + \Pr(F_2) - \Pr(F_1 \cup F_2) = 0.2 + 0.2 - 0.33 = 0.07$$

(d) the probability that at least one of the first and fourth types of flaw is present in the manufactured product
Since F_1 and F_4 are mutually exclusive,

$$\Pr(F_1 \cup F_4) = \Pr(F_1) + \Pr(F_4) = 0.2 + 0.01 = 0.21$$

(e) the probability that the first and fourth types of flaw are present in the manufactured product
Since F_1 and F_4 are mutually exclusive,

$$\Pr(F_1 \cap F_4) = \Pr(\phi) = 0$$

(f) the probability that at least one of the first, second, and fourth types of flaw is present in the manufactured product
From equation (1.20),

$$\Pr(F_1 \cup F_2 \cup F_4) = \Pr(F_1) + \Pr(F_2) + \Pr(F_4) - \Pr(F_1 \cap F_2) - \Pr(F_1 \cap F_4) - \Pr(F_2 \cap F_4) + \Pr(F_1 \cap F_2 \cap F_4)$$

Since F_1 and F_4 are mutually exclusive, $F_1 \cap F_4 = \phi$.
Since F_2 and F_4 are mutually exclusive, $F_2 \cap F_4 = \phi$.
Since $F_1 \cap F_4 = F_2 \cap F_4 = \phi$, $F_1 \cap F_2 \cap F_4 = \phi$.

$$\Pr(F_1 \cup F_2 \cup F_4) = 0.2 + 0.2 + 0.01 - 0.07 - 0 - 0 + 0 = 0.34$$

(g) the probability that the second type of flaw is present in the manufactured product but the third type is not
Note that

$$F_2 = F_2 \cap \Omega = F_2 \cap \left(F_3 \cup F_3^c\right) = \left(F_2 \cap F_3\right) \cup \left(F_2 \cap F_3^c\right)$$

Note also that

$$\left(F_2 \cap F_3\right) \cap \left(F_2 \cap F_3^c\right) = F_2 \cap F_3 \cap F_3^c = F_2 \cap \phi = \phi$$

which implies that $\left(F_2 \cap F_3\right)$ and $\left(F_2 \cap F_3^c\right)$ are mutually exclusive. Therefore,

$$\Pr\left(F_2\right) = \Pr\left(F_2 \cap F_3\right) + \Pr\left(F_2 \cap F_3^c\right), \text{and}$$

$$\Pr\left(F_2 \cap F_3^c\right) = \Pr\left(F_2\right) - \Pr\left(F_2 \cap F_3\right) = 0.2 - 0.06 = 0.14 \ \square$$

1.3 Conditional Probability

In some cases, knowledge of the occurrence of one event may alter the probability of occurrence for some other event. In Example 1.2, customers who purchase the deluxe model may be more likely to choose the orange color. In probability modeling, we use conditional probabilities to study these types of situation.

Let A and B be events defined on a sample space Ω. We refer to $\Pr(A|B)$ as the **conditional probability** of event A given the occurrence of event B, where

$$\Pr\left(A|B\right) = \frac{\Pr\left(A \cap B\right)}{\Pr\left(B\right)}$$

Example 1.2 (continued)

Historical data suggests that 35% of customers order the deluxe model. If a customer orders the deluxe model, then there is a 40% chance that the customer chooses the orange color. What is the probability that a customer orders an orange deluxe item?

Let D denote the event that the customer orders the deluxe model.

$$\Pr\left(E_O \cap D\right) = \Pr\left(E_O | D\right)\Pr\left(D\right) = 0.4\left(0.35\right) = 0.14$$

Historical data also suggests that 10% of customers order a maroon deluxe item. If a customer chooses a deluxe item, then what is the probability that it is maroon?

$$\Pr\left(E_M \,|\, D\right) = \frac{\Pr\left(E_M \cap D\right)}{\Pr\left(D\right)} = \frac{0.1}{0.35} = \frac{2}{7} \quad \square$$

Examples of the first type are more common than the second type. In other words, we typically *use* conditional probabilities rather than *compute* them.

The definition of conditional probability leads to three general rules about conditional probability.

Let A and B be events defined on a sample space Ω.

$$\text{If } A \text{ and } B \text{ are mutually exclusive, then } \Pr(A|B) = 0 \tag{1.22}$$

$$\text{If } B \subseteq A, \text{ then } \Pr(A|B) = 1 \tag{1.23}$$

$$\text{If } A \subseteq B, \text{ then } \Pr(A|B) \geq \Pr(A) \tag{1.24}$$

Statement (1.22) indicates that if A and B are mutually exclusive, then the occurrence of B guarantees that A did not occur. Statement (1.23) indicates that if $B \subseteq A$, then the occurrence of B guarantees the occurrence of A. Statement (1.24) indicates that if $A \subseteq B$, then the occurrence of B cannot reduce the likelihood of A.

The definition of conditional probability also facilitates the consideration of dependency between events.

Let A and B be events defined on a sample space Ω. The events A and B are said to be **independent** if and only if

$$\Pr(A \cap B) = \Pr(A)\Pr(B)$$

If A and B are not independent, they are said to be **dependent**. Let $\{A_1, A_2, \ldots\}$ be a collection of events defined on the sample space. These events are said to be **independent** if and only if for all $\{B_1, B_2, \ldots, B_r\} \subseteq \{A_1, A_2, \ldots\}$

$$\Pr\left(\bigcap_{j=1}^{r} B_j\right) = \prod_{j=1}^{r} \Pr\left(B_j\right)$$

Rather than attempting to prove events are independent, we typically assume independence in order to simplify probability computations involving intersections. In general, if A, B, and C are events defined on a sample space Ω, then to compute $\Pr(A \cap B \cap C)$ we must have knowledge of the other seven probabilities in equation (1.20). But if we assume the three events are independent, then $\Pr(A \cap B \cap C) = \Pr(A)\Pr(B)\Pr(C)$.

Example 1.2 (continued)

Suppose we observe three customer orders. Assuming that customers are independent, what is the probability that the first customer orders a deluxe item, the second customer orders a basic item, and the third customer orders a deluxe item?

Let D_i denote the event that customer i chooses a deluxe item, $i = 1, 2, 3$.

$$\Pr\left(D_1 \cap D_2^c \cap D_3\right) = \Pr\left(D_2^c\right)\Pr\left(D_1\right)\Pr\left(D_3\right) = 0.4\left(1-0.4\right)\left(0.4\right) = 0.096 \;\square$$

Independence is an often misunderstood concept. The independence of two events implies that knowledge of the occurrence of one of the events has no impact on the probability that the other event occurs.

If A and B are independent events defined on a sample space Ω, then

$$\Pr(A|B) = \Pr(A) \tag{1.25}$$

Statements (1.22) and (1.25) imply that **mutually exclusive events are not independent**. The occurrence of one event in a pair of mutually exclusive events clearly implies that the other event does not occur.

One of the most useful results of conditional probability is the **Law of Total Probability**. In its simplest form, the Law of Total Probability uses conditional probabilities associated with the occurrence of one event to determine the probability of some other event.

If A and B are events defined on a sample space Ω, then

$$\Pr(A) = \Pr(A|B)\Pr(B) + \Pr(A|B^c)\Pr(B^c) \tag{1.26}$$

Example 1.3

A certain machine's performance can be characterized by the quality of a key component: 94% of machines with a defective key component will fail, whereas only 1% of machines with a nondefective key component will fail. Note that 4% of machines have a defective key component. What is the probability that the machine will fail?

Let D denote the event that the key component is defective.
Let F denote the event that the machine fails.

$$\Pr(F) = \Pr(F|D)\Pr(D) + \Pr(F|D^c)\Pr(D^c) = 0.94(0.04) + 0.01(1-0.04) = 0.0472 \;\square$$

The Law of Total Probability and the definition of conditional probability can be manipulated into a form more commonly known as **Bayes' Theorem**. Bayes' Theorem is used to investigate the likelihood of the initial event occurring.

If A and B are events defined on a sample space Ω, then

$$\Pr(B|A) = \frac{\Pr(A|B)\Pr(B)}{\Pr(A|B)\Pr(B) + \Pr(A|B^c)\Pr(B^c)} = \frac{\Pr(A|B)\Pr(B)}{\Pr(A)} \quad (1.27)$$

Example 1.3 (continued)

Suppose the machine fails. What is the probability that the key component was defective?

$$\Pr(D|F) = \frac{\Pr(F|D)\Pr(D)}{\Pr(F)} = \frac{0.94(0.04)}{0.0472} = 0.7966 \; \square$$

The Law of Total Probability and Bayes' Theorem can both be extended to the case in which the probability of occurrence of the event of interest can best be described in terms of a partition of the sample space.

If A is an event defined on a sample space Ω and $\{B_1, B_2, \ldots\}$ is a partition of Ω, then

$$\Pr(A) = \sum_{j=1}^{\infty} \Pr(A|B_j)\Pr(B_j)$$

and

$$\Pr(B_j|A) = \frac{\Pr(A|B_j)\Pr(B_j)}{\sum_{i=1}^{\infty} \Pr(A|B_i)\Pr(B_i)} = \frac{\Pr(A|B_j)\Pr(B_j)}{\Pr(A)}$$

for all $j = 1, 2, \ldots$

Example 1.4

An inventory system contains four types of products. Customers order one unit of a product at a time: 20% of customers order the first type of product, 30% the second type, 15% the third type, and 35% the fourth type. Due to the policy used to manage the inventory system, the supplier is out of the first type of product 6% of the time, the second 2% of the time, the third 12% of the time, and the fourth 1% of the time. When a customer orders a product that the inventory system does not have, the order cannot be filled, and the customer takes his or her business elsewhere. What is the probability that an order cannot be filled?

Let T_i denote the event that the customer orders product type i, $i = 1, 2, 3, 4$.

Let S denote the event that the order cannot be filled.

$$\Pr(S) = \sum_{j=1}^{4} \Pr(S|T_j)\Pr(T_j) = 0.06(0.2) + 0.02(0.3) + 0.12(0.15) + 0.01(0.35) = 0.0395$$

Suppose a customer order cannot be filled. What is the probability that the customer ordered the second type of product?

$$\Pr(T_2|S) = \frac{\Pr(S|T_2)\Pr(T_2)}{\Pr(S)} = \frac{0.02(0.3)}{0.0395} = 0.1519 \square$$

Homework Problems

1.1 Random Experiments and Events

(1) Consider the following random experiment. A music company plans to release three new albums next week. One album is by a well-established artist (Artist 1), one is by an artist who has had some success (Artist 2), and one is by a new artist (Artist 3). The company will eventually classify each album as either a "hit" or a "flop."

(a) Define the sample space for this random experiment.

Let H_i denote the event that the album by artist i is a hit, $i = 1, 2, 3$.

(b) Identify the individual outcomes that comprise H_1.

(c) Identify the individual outcomes that comprise $H_1 \cup H_2$.

(d) Identify the individual outcomes that comprise $H_2 \cap H_3$.

(e) Identify the individual outcomes that comprise $H_1^c \cap H_2 \cup H_3$.

Express the following events in terms of H_1, H_2, and H_3.

(f) at least one album is a hit

(g) at least two albums are hits

(h) all albums are hits
(i) only the album by Artist 2 is a hit
(j) at most two albums are hits

(2) Prove equation (1.1).

(3) Prove equation (1.4).

(4) Prove equation (1.7).

(5) Prove statement (1.9).

(6) Prove equation (1.14).

1.2 Probability

(1) Patients visiting a particular doctor have symptoms suggesting that they may have at least one of four ailments. These ailments are referred to as A, B, C, D. Historical data suggests that 25% of patients have ailment A, 70% of patients do not have ailment B, 48% of patients have ailment C, and 35% of patients have ailment D. Note that ailment D is a special case of ailment C. Historical data also suggests that 5% of patients have both ailment A and ailment B, 10% of patients have both ailment B and ailment C, no patient ever has both ailment A and ailment C, and no patient ever has both ailment B and ailment D.

(a) What is the probability that a patient has at least one of ailments A and B?
(b) What is the probability that a patient has at least one of ailments A and D?
(c) What is the probability that a patient has ailment A but not ailment B?
(d) What is the probability that a patient does not have any ailments?

(2) Prove equation (1.16).

(3) Prove statement (1.18).

(4) Consider a unit circle with an inscribed equilateral triangle. A chord is selected at random. What is the probability that the chord is longer than a side of the triangle?

1.3 Conditional Probability

(1) Consider homework problem (1) from section (1.2).

(a) If a patient has ailment B, then what is the probability that the patient does not have ailment A?

(b) If a patient does not have ailment B, then what is the probability that the patient does have ailment A?

(c) Does a patient having ailment D depend on whether or not the patient has ailment A? Explain your answer.

(d) Does a patient having ailment D depend on whether or not the patient has ailment C? Explain your answer.

(2) Consider homework problem (1) from section (1.1). The company's marketing department projects that Artist 1's album has a 90% chance of being a hit, Artist 2's album has a 60% chance of being a hit, and Artist 3's album has a 30% chance of being a hit. The marketing department also believes that the performance ratings of the three albums are independent. Compute the probability of each individual outcome in the sample space.

(3) A cutting tool can only be changed prior to processing of a batch of jobs. A new cutting tool will break during the processing of a batch with probability 0.05, a used cutting tool will break during the processing of a batch with probability 0.15, and an old cutting tool will break during the processing of a batch with probability 0.4. Because of the current tool replacement policy, 50% of batches begin with a new cutting tool, 30% begin with a used cutting tool, and 20% begin with an old cutting tool.

(a) What is the probability that the cutting tool breaks during the processing of a batch of jobs?

(b) Given that the cutting tool breaks during the processing of a batch of jobs, what is the probability that the cutting tool was old prior to processing the batch?

(4) At the beginning of a particular work shift, a machine is in one of three states: operating properly, operating in a minor degraded condition, or operating in a major degraded condition. The machine is operating properly at the beginning of 75% of the work shifts, and operating in a minor degraded condition at the beginning of 20% of the work shifts. Given that the machine is operating properly at the beginning of the work shift, it will fail during the shift with probability 0.01. Given that the machine is operating in a minor degraded condition at the beginning of the work shift, it will fail during the shift with probability 0.25. Given that the machine is operating in a

major degraded condition at the beginning of the work shift, it will fail during the shift with probability 0.5.

(a) What is the probability that the machine fails during the work shift?

(b) Given that the machine fails during the work shift, what is the probability that it was operating in a minor degraded condition at the beginning of the work shift?

(5) Prove statement (1.23).

(6) Prove statement (1.24).

(7) Three contractors are bidding for a project, and the client has randomly selected one to receive the contract. Contractor 1 secretly asks the client to divulge only to him one of his competitors that will not receive the contract. The client is concerned that divulging this information would increase Contractor 1's probability of receiving the contract from one-third to one-half. How would you advise the client?

(8) Customers arriving at the food court in a university's student union are classified as student, faculty member, or other. Let S denote the event that the next customer is a student, and let F denote the event that the next customer is a faculty member. Suppose we begin observing the food court at some point in time. Show that the probability that a student arrives before a faculty member is given by $\Pr(S) / [\Pr(S) + \Pr(F)]$.

(9) A city employs two cab companies: the Green Cab Company and the Blue Cab Company. The current situation is such that 85% of cabs are green. One day a pedestrian was struck by a cab that sped away. A witness identified the cab as a blue cab. So, the pedestrian sued the Blue Cab Company in a United States civil court. When tested by the attorneys in the case, the witness (under similar conditions to the day of the incident) was able to correctly identify the color of the cab 80% of the time. How should the jury in the case rule? Note: This problem is taken from Bennett (1999).

(10) Prove equation (1.25).

(11) Prove equation (1.26).

(12) Prove equation (1.27).

Application: Basic Reliability Theory

Reliability (R) is the probability that a system properly performs its intended function over time when operated in the environment for which it was designed. In basic reliability analysis, we consider the operation of a system over some specified length of time, and we assume that it is operated in its intended environment. In addition, we assume that a system is comprised of a set of components which operate independently. Furthermore, we consider three system structures: (1) series, (2) parallel, (3) everything else.

A **series** system is one in which all components must function properly in order for the system to function. We represent this and all other system structures graphically using **reliability block diagrams**. The conceptual analog for the reliability block diagram is the electrical circuit. The primary difference is that actual physical connection of the components is not necessarily implied. The reliability block diagram for a three-component series system is given in Figure A1.1.

Reliability analysis for an n-component series system proceeds as follows. Let S_j denote the event that component j operates properly, let $r_j = \Pr(S_j)$, let S denote the event that the system operates properly, and note that $R = \Pr(S)$. Note that r_j is referred to as the **component reliability** of component j.

Task 1: Construct an expression for the event S in terms of the events $S_1, S_2, \ldots, S_n$.

Task 2: Construct an expression for R in terms of $r_1, r_2, \ldots, r_n$.

Task 3: Consider a three-component series system having component reliabilities $r_1 = 0.95$, $r_2 = 0.91$, and $r_3 = 0.99$. Compute the system reliability, R.

A **parallel** system is one in which the proper function of at least one component implies system function. The reliability block diagram for a three-component parallel system is given in Figure A1.2.

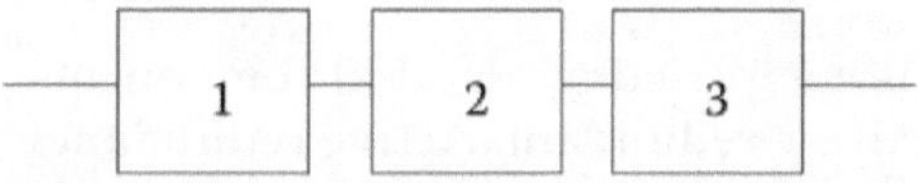

FIGURE A1.1
Three-componet series system

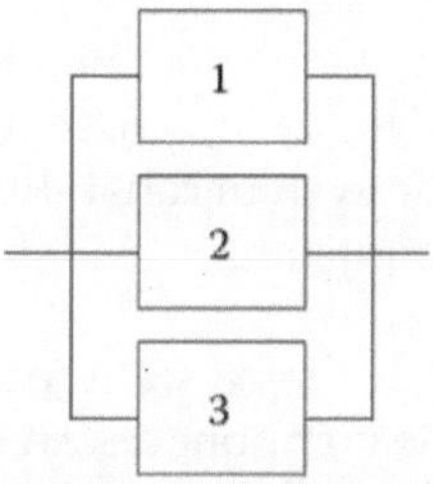

FIGURE A1.2
Three-component parallel system.

Task 4: For an n-component parallel system, construct an expression for the event S in terms of the events $S_1, S_2, \ldots, S_n$.

Task 5: For an n-component parallel system, construct an expression for R in terms of $r_1, r_2, \ldots, r_n$.

Task 6: Simplify the expression constructed in Task 5 for the case in which the n components are identical.

Task 7: Consider a three-component parallel system having component reliabilities $r_1 = 0.95$, $r_2 = 0.91$, and $r_3 = 0.99$. Compute the system reliability, R.

Reliability importance refers to measures used to quantify the impact on system reliability of improving component reliabilities. The most commonly used measure of the importance of component j, as defined by Birnbaum (1969), is

$$I_j = \frac{\partial R}{\partial r_j}$$

Task 8: Construct an expression for the reliability importance of a component in an n-component series system in terms of $r_1, r_2, \ldots, r_n$. Which component in a series system is the most important?

Task 9: Construct an expression for the reliability importance of a component in an n-component parallel system in terms of $r_1, r_2, \ldots, r_n$. Which component in a parallel system is the most important?

It can be shown that any system configuration can be reduced to an equivalent configuration that combines the series and parallel system configurations. In some cases, this reduction results in a loss of component independence. However, we only consider **combined series-parallel** configurations where all components in the system operate independently.

Task 10: Consider the system represented in Figure A1.3. Suppose $r_1 = 0.98$, $r_2 = r_3 = 0.87$, $r_4 = r_5 = 0.94$, $r_6 = 0.99$, $r_7 = r_8 = r_9 = 0.74$. Compute the system reliability, R. Which component in this system is the most important?

If a parallel system (or subsystem) is comprised of identical components, then the components are said to be actively **redundant**. **Active redundancy** is often used in the system design process to make a system more reliable. Unfortunately, redundancy has disadvantages, including increased system cost, increased system weight, etc. Therefore, **redundancy allocation** is used to balance the advantages and disadvantages of redundancy by choosing the number of redundant components to have in each subsystem.

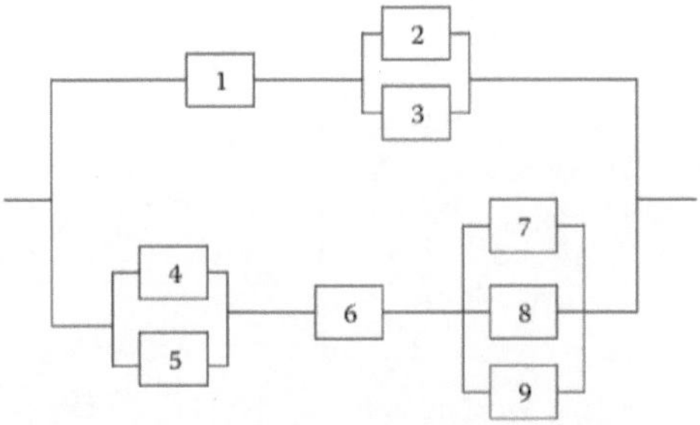

Figure A1.3 Sytstem for task 10.

Task 11: Consider a system comprised of m subsystems connected in series. Let r_i denote the reliability of a single component of subsystem i, let c_i denote the cost of a single component of subsystem i, and let w_i denote the weight of a single component of subsystem i, $i = 1, 2, \dots,$ m. Formulate an optimization model for making redundancy allocation decisions for the system such that system reliability is maximized subject to a system budget of C and a system weight limit of W.

Task 12: Using the model formulated in Task 11, determine the optimal redundancy allocation decisions for the system summarized in Table A1.1. For this system, $C = 48$ and $W = 52$.

TABLE A1.1

Redundancy Allocation Example

i	r_i	c_i	w_i
1	0.92	1	3
2	0.86	2	5
3	0.71	3	4
4	0.88	4	2
5	0.94	5	3

2

Analysis of Random Variables

Random experiments and events serve as the building blocks of all probability models. In engineering applications, however, we are typically interested in quantifying the outcome of a random experiment. If a numerical value is associated with each possible outcome of a random experiment, then we define and use random variables.

2.1 Introduction to Random Variables

We begin with a formal definition of a random variable.

> A **random variable** is a real-valued function defined on a sample space. Random variables are typically denoted by italicized, capital letters. Specific values taken on by a random variable are typically denoted by italicized, lower-case letters.

The random variable of interest depends on the underlying random experiment, its sample space, and the analyst's interest in the random experiment. Common examples of random variables used in industrial engineering applications include dimensions of manufactured products, times required to complete tasks, and demand for goods and services.

Since we are interested in the random variable more than the specific outcomes of the sample space, we need to define probabilities for the possible values of the random variable rather than for events defined on the sample space. The manner in which we assign probabilities to the values taken on by a random variable and the manner in which we answer probability questions regarding a random variable depend on the possible values of the random variable.

> A random variable that can take on at most a countable number of values is said to be a **discrete random variable**. A random variable that can take on an uncountable number of values is said to be a **continuous random variable**. The set of possible values for a random variable is referred to as the **range** of the random variable.

In most industrial engineering applications, random variables that are observed through measurement (part dimensions, elapsed times, etc.) are treated as continuous random variables, and random variables that are observed through counting (number of defects, customer demand, etc.) are treated as discrete random variables.

Regardless of whether a random variable is discrete or continuous, all probability questions regarding a random variable can be answered using its cumulative distribution function.

> If X is a random variable and $F_X(x)$ is a real-valued, nondecreasing, right-continuous function defined for all real x such that
>
> $$F_X(x) = \Pr(X \le x)$$
>
> then F is called the **cumulative distribution function** (CDF) of X. Furthermore,
>
> $$\lim_{x \to \infty} F_X(x) = 1$$
>
> $$\lim_{x \to -\infty} F_X(x) = 0$$
>
> $$\Pr(a < X \le b) = F_X(b) - F_X(a)$$
>
> for all real numbers a and b such that $a < b$, and
>
> $$\Pr(X < c) = \Delta \lim_{\Delta c \to 0+} F_X(c - \Delta c) = F_X(c-)$$
>
> for all real c.

These properties of cumulative distribution functions make it possible to use the CDF to make any probability computation regarding a random variable. One additional point worth noting is that $\Pr(X < c)$ for some real number c is not necessarily equal to $\Pr(X \le c)$ since the latter includes $\Pr(X = c)$.

Example 2.1

Consider a random variable X having CDF F_X. How can the CDF be used to compute the following probabilities?

(a) $\Pr(6 \le X \le 12)$

$$\Pr(6 \le X \le 12) = \Pr(X \le 12) - \Pr(X < 6) = F_X(12) - F_X(6-)$$

(b) $\Pr(X > -14)$

$$\Pr(X > -14) = 1 - \Pr(X \le -14) = 1 - F_X(-14)$$

(c) $\Pr(X \geq 2.7)$

$$\Pr(X \geq 2.7) = 1 - \Pr(X < 2.7) = 1 - F_X(2.7-)$$

(d) $\Pr(-41.2 \leq X < 13)$

$$\Pr(-41.2 \leq X < 13) = \Pr(X < 13) - \Pr(X < -41.2) = F_X(13-) - F_X(-41.2-)$$

(e) $\Pr(7 < X < 25)$

$$\Pr(7 < X < 25) = \Pr(X < 25) - \Pr(X \leq 7) = F_X(25-) - F_X(7) \square$$

2.2 Discrete Random Variables

Since discrete random variables can take on at most a countable number of values, there is a simpler method for computing probabilities for discrete random variables. This method relies on the use of the probability mass function.

If X is a discrete random variable having CDF F_X and range $\{x_1, x_2, \ldots\}$ and f_X is a real-valued function defined for all real x such that

$$f_X(x) = \Pr(X = x)$$

$$f_X(x_j) > 0, \text{ for all } j \in \{1, 2, \ldots\}$$

$$f_X(x) = 0, \text{ if } x \notin \{x_1, x_2, \ldots\}$$

$$\sum_{j=1}^{\infty} f_X(x_j) = 1$$

and

$$F_X(x) = \sum_{x_j \leq x} f_X(x_j)$$

then f_X is called the **probability mass function** (PMF) of X.

Example 2.2

A company owns four service vehicles. Let X denote the number of vehicles in use at some point in time. The discrete random variable X has PMF f_X and CDF F_X, where $f_X(1) = 0.1$, $f_X(2) = 0.4$, $f_X(3) = 0.2$, and $f_X(4) = 0.3$.

(a) Construct the CDF of X.

$$F_X(1) = f_X(1) = 0.1$$

$$F_X(2) = f_X(1) + f_X(2) = 0.5$$

$$F_X(3) = f_X(1) + f_X(2) + f_X(3) = 0.7$$

$$F_X(4) = f_X(1) + f_X(2) + f_X(3) + f_X(4) = 1$$

$$F_X(x) = \begin{cases} 0 & x < 1 \\ 0.1 & 1 \le x < 2 \\ 0.5 & 2 \le x < 3 \\ 0.7 & 3 \le x < 4 \\ 1 & x \ge 4 \end{cases}$$

(b) What is the probability that no more than two vehicles are in use?

$$\Pr(X \le 2) = F_X(2) = f_X(1) + f_X(2) = 0.5$$

(c) What is the probability that not all of the vehicles are in use?

$$\Pr(X < 4) = F_X(4-) = f_X(1) + f_X(2) + f_X(3) = 0.7$$

(d) What is the probability that at least two vehicles are in use?

$$\Pr(X \ge 2) = 1 - \Pr(X < 2) = 1 - F_X(2-) = 1 - f_X(1) = 0.9 \; \square$$

2.3 Continuous Random Variables

The method used for computing probabilities associated with a continuous random variable makes use of the probability density function.

The **probability density function** (PDF) of a continuous random variable X is a real-valued function f_X defined for all real x such that $f_X(x) \ge 0$ for all real x and for all subsets of the real line A

$$\Pr(X \in A) = \int_A f_X(x)\, dx$$

The definition of the PDF also leads to several other results.

If X is a continuous random variable having PDF f_X and CDF F_X and a, b, and c are real numbers such that $a \le b$, then

$$\int_{-\infty}^{\infty} f_X(x)\ dx = 1$$

$$\Pr(a \le X \le b) = \int_{a}^{b} f_X(x)\ dx$$

$$\Pr(X \le c) = \int_{-\infty}^{c} f_X(x)\ dx$$

$$\Pr(X \ge c) = \int_{c}^{\infty} f_X(x)\ dx$$

and

$$\Pr(X = c) = \Pr(c \le X \le c) = \int_{c}^{c} f_X(x)\ dx = 0$$

The relationship between the PDF and the CDF can be summarized by

$$F_X(x) = \int_{-\infty}^{x} f_X(t)\ dt$$

and

$$f_X(x) = \frac{d}{dx} F_X(x)$$

However, these relationships do not provide adequate intuition as to the interpretation of the PDF. For a discrete random variable, the PMF assigns probabilities to the possible values of the random variable. However, the probability of any specific value for a continuous random variable is 0. Therefore, the PDF does not provide the probability of a specific value. Instead, the

PDF provides the relative likelihood (as compared to other possible values) that the random variable will be near a certain value. Therefore, a plot of the PDF provides great insight into the practical range of the random variable and its more likely values.

Because the probability of any specific value of a continuous random variable is 0

$$\Pr(a \le X \le b) = \Pr(a < X \le b) = \Pr(a \le X < b) = \Pr(a < X < b) = F_X(b) - F_X(a)$$

Example 2.3

Let X denote the proportion of patients arriving at an emergency room during a given month who are admitted to the hospital. Suppose X is a continuous random variable having the following PDF where c is a constant:

$$f_X(x) = \begin{cases} c(1-x^2) & 0 \le x \le 1 \\ 0 & \text{otherwise} \end{cases}$$

(a) What is the value of c?

$$\int_{-\infty}^{\infty} f_X(x)\, dx = \int_0^1 c(1-x^2)\, dx = cx - \frac{c}{3}x^3\Big|_0^1 = c - \frac{c}{3} = 1$$

$$c = 1.5$$

(b) Construct the CDF of X.

Let F_X denote the CDF of X.

$$F_X(x) = \int_{-\infty}^{x} f_X(t)\, dt = \int_0^x 1.5 - 1.5t^2\, dt = 1.5t - 0.5t^3\Big|_0^x = 1.5x - 0.5x^3$$

$$F_X(x) = \begin{cases} 0 & x < 0 \\ 1.5x - 0.5x^3 & 0 \le x \le 1 \\ 1 & x > 1 \end{cases}$$

(c) Compute $\Pr(0.2 < X \le 0.8)$.

$$\Pr(0.2 < X \le 0.8) = F_X(0.8) - F_X(0.2)$$

$$\Pr(0.2 < X \leq 0.8) = \left(1.5(0.8) - 0.5(0.8)^3\right) - \left(1.5(0.2) - 0.5(0.2)^3\right)$$

$$\Pr(0.2 < X \leq 0.8) = 0.944 - 0.296 = 0.648$$

(d) Compute Pr(X > 0.5).

$$\Pr(X > 0.5) = 1 - \Pr(X \leq 0.5) = 1 - F_X(0.5) = 0.3125 \quad \square$$

2.4 Expectation

Analysts often choose to summarize probability distributions with specific measures of the behavior of the random variable. The most common of these measures is the expected value. If a random variable X were observed over an infinite number of independent and identical random experiments, then the expected value (mean) of X would be the average of those observations. The method used to compute the expected value of X depends on whether the random variable of interest is discrete or continuous.

Let X be a discrete random variable having PMF f_X and range $\{x_1, x_2, \ldots\}$. The **expected value** or **mean** of X is denoted by E(X) and given by

$$E(X) = \sum_{j=1}^{\infty} x_j f_X(x_j)$$

Let X be a continuous random variable having PDF f_X. The **expected value** of X is denoted by E(X) and given by

$$E(X) = \int_{-\infty}^{\infty} x \; f_X(x) \; dx$$

The expected value of X is a weighted average of the possible values of X. The weights, in the discrete case, are determined by the probabilities of the possible values. In the continuous case, the weight assigned to each possible value is the corresponding value of the PDF.

Example 2.2 (continued)

Determine the expected value of X.

$$E(X) = 1(0.1) + 2(0.4) + 3(0.2) + 4(0.3) = 2.7 \text{ vehicles} \ \square$$

Example 2.3 (continued)

Determine the expected value of X.

$$E(X) = \int_0^1 1.5x - 1.5x^3 \, dx = 0.375 \ \square$$

In some cases, we may also be interested in the expected value of a function of the random variable. In Example 2.2, we may be interested in the expected value of the fuel consumption of the four vehicles. In such cases, we do not have to determine the probability distribution of the function of interest in order to determine its expected value.

Let X be a discrete random variable having PMF f_X and range $\{x_1, x_2, \ldots\}$. If g is a real-valued **function of the random variable** defined on the range of X, then

$$E[g(X)] = \sum_{j=1}^{\infty} g(x_j) f_X(x_j)$$

Let X be a continuous random variable having PDF f_X. If g is a real-valued **function of the random variable** defined for all real x, then

$$E[g(X)] = \int_{-\infty}^{\infty} g(x) f_X(x) dx$$

The reference to the function g as a function of a random variable is somewhat misleading, because g is in fact a random variable. This method of computing the expected value of a function of a random variable is sometimes referred to as the **Law of the Unconscious Statistician** because it allows the computation of the expected value of $g(X)$ without determining the probability distribution of $g(X)$.

Regardless of whether the random variable is discrete or continuous, there is one case in which it is quite easy to compute the expected value of a function of the random variable.

If X is a random variable and a and b are constants, then

$$E(aX + b) = aE(X) + b \tag{2.1}$$

Whereas the expected value measures the central tendency of a random variable, other measures characterize the dispersion of the probability distribution. The two most common of these measures are the variance and standard deviation of the random variable.

Let X be a random variable. The **variance** of X is denoted by $Var(X)$ and given by

$$Var(X) = E\{[X - E(X)]^2\} = E(X^2) - [E(X)]^2 \tag{2.2}$$

The **standard deviation** of X, denoted by $StDev(X)$, is the positive square root of $Var(X)$.

The variance and standard deviation are both non-negative quantities, but the standard deviation is typically preferred over variance as a descriptor of random variable behavior. This preference is a result of the fact that the standard deviation has the same units of measure as the random variable itself.

If a random variable X were observed over an infinite number of independent and identical random experiments, then the standard deviation of X would be the average absolute deviation of each observation from the expected value of X. Therefore, a larger standard deviation implies more variability in the behavior of the random variable.

Example 2.2 (continued)

Determine the variance of X.

$$E(X^2) = 1^2(0.1) + 2^2(0.4) + 3^2(0.2) + 4^2(0.3) = 8.3 \text{ vehicles}^2$$

$$Var(X) = 8.3 - (2.7)^2 = 1.01 \text{ vehicles}^2 \ \square$$

Example 2.3 (continued)

Determine the standard deviation of X.

$$E(X^2) = \int_0^1 1.5x^2 - 1.5x^4 \, dx = 0.2$$

$$Var(X) = 0.2 - (0.375)^2 = 0.059375$$

$$StDev(X) = \sqrt{0.059375} = 0.2437 \ \square$$

As with the expected value, we can also compute the variance of a function of a random variable. In the linear case, this computation is again straightforward.

If X is a random variable and a and b are constants, then

$$Var(aX+b)=a^2Var(X) \tag{2.3}$$

Example 2.4

A store sells a certain type of component in lots of 100. When an order is placed, the customer can also order spare components. Historical data suggests that the number of spares ordered by a single customer is one of the integers in the set {0, 1, ... , 4} and that these five integers all have the same probability. The lot of 100 and the spares are shipped in a single container. Let X denote the number of spares ordered. The shipping cost for an order is \$10 plus \$2 for each spare component. Let Y denote the shipping cost for a single order.

(a) Determine the expected value and standard deviation of X.

$$E(X)=\sum_{x=0}^{4}x\left(\frac{1}{5}\right)=2 \text{ components}$$

$$E(X^2)=\sum_{x=0}^{4}x^2\left(\frac{1}{5}\right)=6 \text{ components}^2$$

$$Var(X)=6-(2)^2=2 \text{ components}^2$$

$$StDev(X)=\sqrt{2}=1.4142 \text{ components}$$

(b) Determine the expected value and standard deviation of Y.

$$Y=\$2X+\$10$$

$$E(Y)=2E(X)+10=\$14$$

$$Var(Y)=2^2Var(X)=8 \text{ dollars}^2$$

$$StDev(Y)=\sqrt{8}=\$2.83 \ \square$$

2.5 Generating Functions

The Law of the Unconscious Statistician also can be used to obtain two other useful functions in analyzing the behavior of random variables: the moment generating function and the probability generating function.

> If X is a random variable, then $E(X^n)$ is referred to as the ***n*th moment** of X. The **moment generating function** of X is denoted by $M_X(s)$ and given by
>
> $$M_X(s) = E(e^{sX})$$

The expected value of X is the 1st moment of X, and the 2nd moment of X is used in computing the variance of X. As does the CDF, the PMF (in the discrete case), and the PDF (in the continuous case), the moment generating function uniquely determines the probability distribution of a random variable.

As implied by its name, the moment generating function of a random variable can be used to compute its moments.

> If X is a random variable and $M_X(s)$ denotes the moment generating function of X, then
>
> $$\left.\frac{d^n}{ds^n} M_X(s)\right|_{s=0} = E(X^n) \qquad (2.4)$$
>
> $n = 1, 2, \ldots$

Example 2.5

Consider an emergency services dispatching telephone center. Let X denote the number of calls received in one month by the center. Suppose the range of X is $\{0, 1, \ldots\}$, and let f_X denote the PMF of X where

$$f_X(x) = \frac{e^{-350}350^x}{x!}, x = 0, 1, \ldots$$

(a) Derive the moment generating function of X.

$$M_X(s) = E(e^{sX}) = \sum_{x=0}^{\infty} e^{sx}\frac{e^{-350}350^x}{x!} = e^{-350}\sum_{x=0}^{\infty}\frac{(350e^s)^x}{x!}$$

$$M_X(s) = e^{-350}e^{350e^s} = e^{350(e^s-1)}$$

(b) Use the moment generating function to determine the mean and standard deviation of X.

$$E(X)=\frac{d}{ds}M_X(s)\bigg|_{s=0}=e^{350(e^s-1)}\left(350e^s\right)\bigg|_{s=0}=350 \text{ calls}$$

$$E(X^2)=\frac{d^2}{ds^2}M_X(s)\bigg|_{s=0}=e^{350(e^s-1)}\left(350e^s\right)+\left(350e^s\right)^2 e^{350(e^s-1)}\bigg|_{s=0}=350+(350)^2 \text{ calls}^2$$

$$Var(X)=350+(350)^2-(350)^2=350 \text{ calls}^2$$

$$StDev(X)=\sqrt{350}=18.71 \text{ calls} \ \square$$

Probability generating functions can only be defined for non-negative, discrete random variables.

Let X be a discrete random variable having a range that is a subset of {0, 1, … }. The **probability generating function** of X is denoted by $A_X(s)$ and given by

$$A_X(s)=E\left(s^X\right)$$

As its name implies, the probability generating function can be used to determine probabilities of the individual values of a non-negative, discrete random variable. But in addition, the probability generating function can also be used to indirectly determine the moments of the random variable.

Let X be a discrete random variable having a range that is a subset of {0, 1, … } and probability generating function $A_X(s)$. Then

$$\frac{d^n}{ds^n}A_X(s)\bigg|_{s=1}=E\left[\prod_{i=1}^{n}(X-i+1)\right] \tag{2.5}$$

$n = 1, 2, \ldots$, and

$$\frac{1}{n!}\frac{d^n}{ds^n}A_X(s)\bigg|_{s=0}=\Pr(X=n) \tag{2.6}$$

$n = 0, 1, \ldots$

Example 2.6

Consider a manufacturing process. Each item manufactured is classified as defective or nondefective. Let X denote the number of nondefective items manufactured before the first defective item. Suppose the range of X is {0, 1, … }, and let f_X denote the PMF of X where

$$f_X(x) = 0.04(0.96)^x, x = 0, 1, \ldots$$

(a) Derive the probability generating function of X.

$$A_X(s) = E(s^X) = \sum_{x=0}^{\infty} s^x (0.04)(0.96)^x = 0.04 \sum_{x=0}^{\infty} (0.96s)^x = \frac{0.04}{1 - 0.96s}$$

This function is only defined for $0 \leq 0.96s < 1$.

(b) Use the probability generating function to determine Pr(X = 1) and Pr(X = 2).

$$\Pr(X=1) = \frac{1}{1!}\frac{d}{ds}A_X(s)\bigg|_{s=0} = \frac{0.04(0.96)}{(1-0.96s)^2}\bigg|_{s=0} = 0.04(0.96)$$

$$\Pr(X=2) = \frac{1}{2!}\frac{d^2}{ds^2}A_X(s)\bigg|_{s=0} = \frac{2(0.04)(0.96)^2}{2(1-0.96s)^3}\bigg|_{s=0} = 0.04(0.96)^2 \quad \square$$

2.6 Common Applications of Random Variables

We conclude this chapter by considering some applications of random variables that are well known but not considered further in this text.

2.6.1 Equally Likely Alternatives

Our first application considers random variables such that each value in the range is equally likely to occur.

Let X be a discrete random variable having a range corresponding to a finite set of consecutive integers $\{a, a + 1, \ldots, b\}$. If each value in the range has the same probability, then X is said to be a **discrete uniform random variable.** This fact is denoted by $X \sim DU(a,b)$. Furthermore, the PMF of X is given by

$$f_X(x) = \frac{1}{b-a+1}, x = a, a+1, \ldots, b$$

$$E(X) = \frac{a+b}{2}$$

and

$$Var(X) = \frac{(b-a+1)^2 - 1}{12}$$

Example 2.7

A university bookstore sells party tents having the university logo on the canopy. The number of tents sold by the bookstore during the month of September is a discrete uniform random variable having range $\{20, 21, \ldots, 40\}$.

(a) What is the probability that the bookstore sells 25 tents during the month of September?

If X denotes the number of tents sold by the bookstore during the month of September, then $X \sim DU(20,40)$. Let f_X denote the PMF of X.

$$\Pr(X = 25) = f_X(25) = \frac{1}{40-20+1} = \frac{1}{21}$$

(b) Determine the probability that the bookstore sells more than 30 tents during the month of September.

$$\Pr(X > 30) = \sum_{x=31}^{40} f_X(x) = \sum_{x=31}^{40} \frac{1}{40-20+1} = \frac{1}{21}\sum_{x=31}^{40} 1 = \frac{10}{21}$$

(c) Determine the expected value and the standard deviation of the number of tents sold by the bookstore during the month of September.

$$E(X) = \frac{20+40}{2} = 30 \text{ tents}$$

$$StDev(X) = \sqrt{Var(X)} = \sqrt{\frac{(40-20+1)^2-1}{12}} = 6.0553 \text{ tents} \ \square$$

A continuous version of equally likely alternatives is also well known.

Let X be a continuous random variable having range $[a,b]$, and suppose all the values within the range of X are equally likely. The PDF of X is given by

$$f_X(x) = \begin{cases} \frac{1}{b-a} & a \le x \le b \\ 0 & \text{otherwise} \end{cases}$$

and X is said to be a (continuous) **uniform random variable**. This fact is denoted by $X \sim U(a,b)$. Furthermore, the CDF of X is given by

$$F_X(x) = \begin{cases} 0 & x < a \\ \frac{x-a}{b-a} & a \le x \le b \\ 1 & x > b \end{cases}$$

$$E(X) = \frac{a+b}{2}$$

$$Var(X) = \frac{(b-a)^2}{12}$$

and

$$M_X(s) = \frac{e^{sb} - e^{sa}}{s(b-a)}$$

Example 2.8

Let X denote the temperature in the accelerated testing chamber in an electronics manufacturing facility. Suppose X is uniformly distributed over the interval (110°F, 130°F).

(a) Identify the PDF and CDF of X.

Note that $X \sim U(110,130)$ where X is measured in degrees Fahrenheit. Let f_X denote the PDF of X, and let F_X denote the CDF of X.

$$f_X(x) = \begin{cases} \dfrac{1}{130-110} & 110 \le x \le 130 \\ 0 & \text{otherwise} \end{cases}$$

$$F_X(x) = \begin{cases} 0 & x < 110 \\ \dfrac{x-110}{130-110} & 110 \le x \le 130 \\ 1 & x > 130 \end{cases}$$

(b) Determine the expected value and standard deviation of X.

$$E(X) = \frac{110+130}{2} = 120^\circ\text{F}$$

$$StDev(X) = \sqrt{Var(X)} = \sqrt{\frac{(130-110)^2}{12}} = 5.77^\circ\text{F}$$

(c) Determine the probability that the temperature in the chamber exceeds 125°F.

$$\Pr(X > 125) = 1 - F_X(125) = 1 - \frac{125-110}{130-110} = 0.25$$

(d) Determine the value of x such that $\Pr(X < x) = 0.1$.

$$\Pr(X < x) = F_X(x) = \frac{x-110}{130-110} = 0.1$$

$$x = 0.1(130-110) + 110 = 112^\circ\text{F} \ \square$$

Monte Carlo and discrete-event simulation both depend on the use of observations drawn from a uniform probability distribution. These observations are referred to as random numbers.

If $X \sim U(0,1)$, then an observation of X is referred to as a **random number.**

2.6.2 Random Sampling

We now consider two scenarios that address the sampling and classification of items from a population.

Consider a population of N items of which K belong to a classification of interest. Let $p = K/N$. Suppose n items are selected at random from the population in a sequential fashion. After each item is sampled, its classification is noted and then it is returned to the population. This process is referred to as **sampling with replacement.**

If X denotes the number of sampled items that belong to the classification of interest, then X is said to be a **binomial random variable**. This fact is denoted by $X \sim \text{bin}(n,p)$, and the range of X is $\{0, 1, \ldots, n\}$. Furthermore, the PMF of X is given by

$$f_X(x) = \binom{n}{x} p^x (1-p)^{n-x} = \frac{n!}{x!(n-x)!} p^x (1-p)^{n-x}$$

$x = 0, 1, \ldots, n,$

$$E(X) = np$$

$$Var(X) = np(1 - p)$$

$$M_X(s) = (1 - p + pe^s)^n$$

and

$$A_X(s) = (1 - p + ps)^n$$

The second scenario is identical to the first except that sampled items are not returned to the population.

Consider a population of N items of which K belong to a classification of interest. Let $p = K/N$. Suppose n items are selected at random from the population in a sequential fashion. After each item is sampled, its classification is noted and it is not returned to the population. This process is referred to as **sampling without replacement**.

If X denotes the number of sampled items that belong to the classification of interest, then X is said to be a **hypergeometric random variable**.

This fact is denoted by $X \sim HG(N,K,n)$, and the range of X is $\{x_{min}, x_{min}+1, \ldots, x_{max}\}$ where $x_{min} = \max(0, n - N + K)$ and $x_{max} = \min(n, K)$. Furthermore, the probability mass function of X is given by

$$f_X(x) = \frac{\binom{K}{x}\binom{N-K}{n-x}}{\binom{N}{n}}$$

for any value x in the range of X,

$$E(X) = np$$

and

$$Var(X) = np(1-p)\frac{N-n}{N-1}$$

Example 2.9

In a shipment of 100 items, 12 are defective and 88 are not defective. A random sample of 10 items is taken from the shipment.

(a) If the sample is taken with replacement, then what is the probability that three of the sampled items are defective?

Let X denote the number of sampled items that are defective. Note that $X \sim \text{bin}(10,p)$ where

$$p = \frac{12}{100}$$

Let f_X denote the probability mass function of X.

$$\Pr(X=3) = f_X(3) = \binom{10}{3}(0.12)^3(0.88)^{10-3} = \frac{10!}{3!\ 7!}(0.12)^3(0.88)^7 = 0.0847$$

(b) If the sample is taken without replacement, then what is the probability that three of the sampled items are defective?

Let Y denote the number of sampled items that are defective. Note that $Y \sim$ HG(100, 12, 10). Let g_Y denote the probability mass function of X.

$$\Pr(Y=3) = g_Y(3) = \frac{\binom{12}{3}\binom{88}{7}}{\binom{100}{10}} = 0.0807 \quad \square$$

2.6.3 Normal Random Variables

The normal random variable is often used to represent physical quantities that have a nominal (expected) value of μ but are subject to random noise (which is captured by the standard deviation σ). The "bell curve" shape of the normal PDF is widely recognized even by many people with little experience in probability. The normal random variable also has numerous applications in statistical analysis.

If X is a **normal random variable** having mean μ and standard deviation σ, denoted by $X \sim N(\mu,\sigma^2)$, then the PDF of X is given by

$$f_X(x) = \frac{1}{\sqrt{2\pi}\sigma} e^{-(x-\mu)^2/2\sigma^2}$$

for all real x,

$$E(X) = \mu$$

$$Var(X) = \sigma^2$$

and

$$M_X(s) = e^{\mu s + \frac{\sigma^2 s^2}{2}}$$

There is no closed-form expression for the CDF of a normal random variable. To avoid any significant complications resulting from this fact, we make use of the standard normal random variable.

Denoted by Z, the **standard normal random variable** is a normal random variable having a mean of 0 and a standard deviation of 1. Furthermore, if $X \sim N(\mu,\sigma^2)$, then

$$Z = \frac{X - \mu}{\sigma}$$

is a standard normal random variable.

CDF values of the standard normal random variable are tabulated in almost all books having significant discussion of statistical analysis.

Example 2.10

The width of a manufactured part is a normal random variable having a mean of 4 cm and a standard deviation of 0.1 cm. What is the probability that a part has a width greater than 4.15 cm?

Let X denote the width of the part. Note that $X \sim N(4, 0.1^2)$ where X is measured in centimeters.

$$\Pr(X > 4.15) = \Pr\left(\frac{X-4}{0.1} > \frac{4.15-4}{0.1}\right) = \Pr(Z > 1.5) = 1 - \Pr(Z \le 1.5)$$

$$\Pr(X > 4.15) = 1 - 0.9332 = 0.0668 \quad \square$$

Homework Problems

2.2 Discrete Random Variables

(1) A delivery truck is required to make five stops on its morning route. Let X denote the number of on-time deliveries for a given morning, let f_X denote the PMF of X, and let F_X denote the CDF of X. Historical data suggests that

$$f_X(5) = 0.88, f_X(4) = 0.06, f_X(3) = 0.03, f_X(2) = 0.02, f_X(1) = 0.008, f_X(0) = 0.002$$

(a) Construct the CDF of X.

(b) Determine the probability that the truck makes at least four on-time deliveries.

2.3 Continuous Random Variables

(1) Let X denote the life length of a mechanical component (measured in days), and let f_X denote the PDF of X where

$$f_X(x) = \begin{cases} \frac{1}{800} x \exp\left(\frac{-x^2}{1600}\right) & x > 0 \\ 0 & \text{otherwise} \end{cases}$$

(a) Derive the CDF of X.

(b) Determine the probability that the component survives more than 30 days.

(c) Determine the probability that the component survives more than 30 days but less than 50 days.

2.4 Expectation

(1) Consider the random variable X defined in problem (1) under 2.2 Discrete Random Variables above.

(a) Determine the expected value of X.

(b) Determine the standard deviation of X.

(c) Suppose the driver is paid \$150 per day, but is charged a \$30 deduction for each late delivery. Let Y denote the amount the driver is paid for one day. Determine the expected value and standard deviation of Y.

(2) A store sells a variety of items. Let X denote the weekly demand for an item. The range of X is $\{0, 1, \ldots\}$ and the PMF of X is given by

$$f_X(x) = \frac{e^{-\alpha}\alpha^x}{x!}, x = 0, 1, \ldots$$

where α is a constant such that $\alpha > 0$. Determine the expected value and standard deviation of X.

(3) Let X denote the time between customer arrivals at a bank, and let f_X denote the PDF of X where X is measured in minutes and

$$f_X(x) = \begin{cases} 0.2e^{-0.2x} & x > 0 \\ 0 & x \le 0 \end{cases}$$

Determine the expected value and standard deviation of X.

(4) Prove that the second equality in equation (2.2) holds for a discrete random variable X having range $\{x_1, x_2, \dots\}$.

(5) Prove that equation (2.1) holds for a discrete random variable X having range $\{x_1, x_2, \dots\}$.

(6) Prove that equation (2.3) holds for a continuous random variable X.

2.5 Generating Functions

(1) Consider the random variable X defined in Example 2.6.
  (a) Derive the moment generating function of X.
  (b) Use the moment generating function to determine the expected value of X and the standard deviation of X.

(2) Prove that equation (2.4) holds for a continuous random variable X.

(3) Consider the random variable X defined in problem (3) under 2.4 Expectation above.
  (a) Derive the moment generating function of X.
  (b) Use the moment generating function to determine the expected value of X and the standard deviation of X.

(4) Consider the random variable X defined in Example 2.5.
  (a) Derive the probability generating function of X.
  (b) Use the probability generating function to determine $\Pr(X = 0)$, $\Pr(X = 1)$, and $\Pr(X = 2)$.
  (c) Use the probability generating function to determine the expected value of X and the standard deviation of X.

(5) Prove that equations (2.5) and (2.6) hold for a discrete random variable X having range $\{0, 1, \dots\}$.

2.6 Common Applications of Random Variables

(1) A room contains 15 people. What is the probability that all 15 people have unique birthdays (considering month and day only)? Clearly state your assumptions.

Application: Basic Warranty Modeling

In traditional models of product failure, reliability is determined using a **time to failure** probability distribution. Let T denote the time to failure for a product, let F_T denote the CDF of T, and suppose that

$$F_T(t) = \begin{cases} 0 & t \leq 0 \\ 1 - \exp\left[-\left(\dfrac{t}{\eta}\right)^{\beta}\right] & t > 0 \end{cases}$$

where $\beta > 1$ and $\eta > 0$. Under such a model, T is said to have a Weibull distribution.

Task 1: Let f_T denote the PDF of T. Derive $f_T(t)$ and $E(T)$.

When reliability is determined using a time to failure distribution, reliability is defined as the probability that the product survives more than t time units and denoted by

$$R(t) = 1 - F_T(t)$$

Suppose the sale of the product includes the following warranty.

> If the product fails within τ time units, the customer will be refunded the purchase price s.

Not all customers take advantage of the warranty. Let a denote the probability that an eligible customer uses the warranty, and let c_a denote the cost of processing a warranty claim.

Task 2: Let $W(T,\tau)$ denote the warranty cost for a product that fails at time T under a warranty policy τ. Construct an expression for $E[W(T,\tau)]$ in terms of $F_T(t)$.

The failure of a product also implies some lost goodwill on the part of the customer. Let $G(T,\tau)$ denote the cost of lost goodwill if failure occurs at time T under a warranty policy τ where

$$G(T,\tau)=\begin{cases} g_0 & T \le \tau \\ g_1 & \tau < T \le t_{\max} \\ 0 & T > t_{\max} \end{cases}$$

and $t_{\max}$ is an upper bound on the amount of time a customer intends to use the product.

Task 3: Construct an expression for $E[G(T,\tau)]$ in terms of $F_T(t)$.

Task 4: Let $P(T,\tau)$ denote the profit generated by a product that fails at time t under a warranty policy τ. Construct an expression for $E[P(T,\tau)]$ in terms of $E[W(T,\tau)]$ and $E[G(T,\tau)]$.

The choice of the values of τ impacts the sales for a product (a better warranty leads to increased sales). Let $d(\tau)$ denote the demand for a product resulting from a warranty policy τ where

$$d(\tau)=d_0+(d_1-d_0)\left(1-e^{-k\tau}\right)$$

d_0 denotes the minimum demand, d_1 denotes the maximum demand, and $k > 0$ is a parameter.

Task 5: Let $B(\tau)$ denote the total profit resulting from all sales if a warranty policy τ is implemented. Construct an expression for $E[B(\tau)]$ in terms of $d(\tau)$ and $E[P(T,\tau)]$.

Suppose the parameter values for a particular product are:

$$\beta = 2.2,\ \eta = 100 \text{ weeks}$$

$$s = \$100,\ a = 0.7,\ c_a = \$8$$

$$g_0 = \$5,\ g_1 = \$12,\ t_{\max} = 250 \text{ weeks}$$

$$d_0 = 15{,}000 \text{ units},\ d_1 = 30{,}000 \text{ units},\ k = 0.02$$

Task 6: Recommend a warranty policy τ to the manufacturer. Support this recommendation with numerical results.

Task 7: Of the product parameters, g_0 and g_1 are the most difficult to estimate. Evaluate the sensitivity of the recommended warranty policy to these values.

3

Analysis of Multiple Random Variables

In Chapter 2, we considered issues related to a single random variable defined on a random experiment. In some cases, we are interested in more than one random variable. For example, a manufactured product may have more than one measurable quality characteristic. In this chapter, we consider situations in which there are two random variables of interest. However, all of the concepts explored in this chapter can be extended to more than two random variables.

3.1 Two Random Variables

As with a single random variable, the study of two random variables may begin with the definition of a cumulative distribution function. Consider two random variables, X and Y, defined on a single random experiment. In other words, each outcome of the random experiment has an associated value of X and Y.

If $F_{X,Y}$ is a function such that

$$F_{X,Y}(x, y) = \Pr(X \leq x, Y \leq y)$$

for all real x and all real y, then $F_{X,Y}$ is referred to as the **joint cumulative distribution function** (joint CDF) of X and Y. The function $F_{X,Y}$ is nondecreasing in x, nondecreasing in y, right-continuous with respect to x, and right-continuous with respect to y.

We can use the joint CDF to determine the individual (marginal) CDFs for the two random variables.

Consider two random variables, X and Y, defined on a single random experiment. If $F_{X,Y}$ denotes the joint CDF of X and Y, F_X denotes the **marginal cumulative distribution function** of X, and F_Y denotes the marginal cumulative distribution function of Y, then

$$F_X(x) = \lim_{y \to \infty} F_{x,y}(x, y)$$

and

$$F_Y(y) = \lim_{x \to \infty} F_{x,y}(x, y)$$

3.1.1 Two Discrete Random Variables

As with the study of single random variables, the analysis of multiple random variables is simplified by classifying the random variables as discrete or continuous. First, we consider the case in which the two random variables are both discrete.

Let X and Y be discrete random variables defined on a single random experiment such that $\{x_1, x_2, \ldots\}$ is the range of X and $\{y_1, y_2, \ldots\}$ is the range of Y. Let $F_{X,Y}$ denote the joint CDF of X and Y. If $f_{X,Y}$ is a function such that

$$f_{X,Y}(x, y) = \Pr(X = x, Y = y)$$

for all real x and all real y, then $f_{X,Y}$ is referred to as the **joint probability mass function** (joint PMF) of X and Y. Note that

$$f_{X,Y}(x, y) \geq 0$$

for all real x and all real y,

$$f_{X,Y}(x, y) = 0$$

if $x \notin \{x_1, x_2, \ldots\}$ or $y \notin \{y_1, y_2, \ldots\}$,

$$\sum_{j=1}^{\infty} \sum_{k=1}^{\infty} f_{X,Y}(x_j, y_k) = 1$$

and

$$F_{X,Y}(x, y) = \sum_{x_j \leq x} \sum_{y_k \leq y} f_{X,Y}(x_j, y_k)$$

As with the joint CDF, we can use the joint PMF to find the PMFs for the individual random variables.

Let X and Y be discrete random variables defined on a single random experiment such that $\{x_1, x_2, \ldots\}$ is the range of X and $\{y_1, y_2, \ldots\}$ is the range of Y. If $f_{X,Y}$ denotes the joint PMF of X and Y, f_X denotes the

marginal probability mass function of X, and f_Y denotes the marginal probability mass function of Y, then

$$f_X(x_j) = \sum_{k=1}^{\infty} f_{X,Y}(x_j, y_k)$$

$j = 1, 2, \dots$, and

$$f_Y(y_k) = \sum_{j=1}^{\infty} f_{X,Y}(x_j, y_k)$$

$k = 1, 2, \dots$.

Example 3.1

A service contractor provides equipment maintenance services to three customers (customer 1, customer 2, customer 3). When a service call is received by the contractor, the dispatcher records the customer number (1, 2, or 3) and the number of units of equipment in need of maintenance. Let X denote the customer number, and let Y denote the number of units of equipment in need of maintenance. Let $f_{X,Y}$ denote the joint PMF of X and Y where

$f_{X,Y}(1,1) = 0.18$	$f_{X,Y}(1,2) = 0.11$	$f_{X,Y}(1,3) = 0.06$	$f_{X,Y}(1,4) = 0.04$
$f_{X,Y}(2,1) = 0.15$	$f_{X,Y}(2,2) = 0.12$	$f_{X,Y}(2,3) = 0.07$	$f_{X,Y}(2,4) = 0.05$
$f_{X,Y}(3,1) = 0.09$	$f_{X,Y}(3,2) = 0.07$	$f_{X,Y}(3,3) = 0.05$	$f_{X,Y}(3,4) = 0.01$

(a) Determine the probability that the call is for either customer 1 or 2 and involves more than two units of equipment.

$$\Pr(X \le 2, Y > 2) = f_{X,Y}(1,3) + f_{X,Y}(1,4) + f_{X,Y}(2,3) + f_{X,Y}(2,4) = 0.22$$

(b) Determine the marginal PMF of X.

Let f_X denote the marginal PMF of X.

$$f_X(1) = \sum_{y=1}^{4} f_{X,Y}(1, y) = 0.39$$

$$f_X(2) = \sum_{y=1}^{4} f_{X,Y}(2, y) = 0.39$$

$$f_X(3) = \sum_{y=1}^{4} f_{X,Y}(3, y) = 0.22$$

(c) Determine the marginal PMF of Y.

Let f_Y denote the marginal probability mass function of Y.

$$f_Y(1) = \sum_{x=1}^{3} f_{X,Y}(x, 1) = 0.42$$

$$f_Y(2) = \sum_{x=1}^{3} f_{X,Y}(x, 2) = 0.30$$

$$f_Y(3) = \sum_{x=1}^{3} f_{X,Y}(x, 3) = 0.18$$

$$f_Y(4) = \sum_{x=1}^{3} f_{X,Y}(x, 4) = 0.1 \quad \square$$

3.1.2 Two Continuous Random Variables

Our treatment of two continuous random variables begins with the definition of a joint probability density function.

> The **joint probability density function** (joint PDF) for two continuous random variables, X and Y, defined on a single random experiment is a real-valued function $f_{X,Y}$ such that
>
> $$f_{X,Y}(x,y) \geq 0$$
>
> for all real x and all real y, and for all subsets of the real line A and all subsets of the real line B
>
> $$\Pr(X \in A, Y \in B) = \int_A \int_B f_{X,Y}(x, y)\, dy\, dx$$
>
> Note that
>
> $$\int_{-\infty}^{\infty} \int_{-\infty}^{\infty} f_{X,Y}(x, y)\, dy\, dx = 1.$$

The relationship between the joint CDF and the joint PDF is analogous to the case of a single, continuous random variable.

> Let X and Y be continuous random variables defined on a single random experiment. If $f_{X,Y}$ denotes the joint PDF of X and Y, and $F_{X,Y}$ denotes the joint CDF of X and Y, then
>
> $$F_{X,Y}(x,y) = \int_{-\infty}^{x}\int_{-\infty}^{y} f_{X,Y}(t,u)\,du\,dt$$
>
> and
>
> $$f_{X,Y}(x,y) = \frac{d^2}{dx\,dy} F_{X,y}(x,y)$$

We can use the joint PDF to obtain the individual (marginal) PDFs.

> Let X and Y be continuous random variables defined on a single random experiment. If $f_{X,Y}$ denotes the joint PDF of X and Y, f_X denotes the **marginal probability density function** of X, and f_Y denotes the marginal probability density function of Y, then
>
> $$f_X(x) = \int_{-\infty}^{\infty} f_{X,y}(x,y)\,dy$$
>
> and
>
> $$f_Y(y) = \int_{-\infty}^{\infty} f_{X,Y}(x,y)\,dx$$

Example 3.2

A small production line is comprised of two machines (machine 1 and machine 2). Let X denote the time until the next failure of machine 1, and let Y denote the time until the next failure of machine 2. Both X and Y are measured in days. The joint PDF of X and Y is given by

$$f_{X,Y}(x,y) = \begin{cases} 0.002e^{-(0.1x+0.02y)} & x > 0, y > 0 \\ 0 & \text{otherwise} \end{cases}$$

(a) Construct the joint CDF of X and Y.

Let $F_{X,Y}$ denote the joint CDF of X and Y.

$$F_{X,Y}(x,y) = \int_0^x \int_0^y 0.002e^{-(0.1u+0.02v)} \, dv \, du$$

$$F_{X,Y}(x,y) = 1 - e^{-0.1x} - e^{-0.02y} + e^{-(0.1x+0.02y)}$$

$$F_{X,Y}(x,y) = \begin{cases} 1 - e^{-0.1x} - e^{-0.02y} + e^{-(0.1x+0.02y)} & x > 0, y > 0 \\ 0 & \text{otherwise} \end{cases}$$

(b) Compute the probability that both machines fail within two days.

$$\Pr(X \le 2, Y \le 2) = F_{X,Y}(2,2) = 1 - e^{-0.1(2)} - e^{-0.02(2)} + e^{-(0.1(2)+0.02(2))} = 0.0071$$

(c) Determine the marginal PDF of X.

Let f_X denote the PDF of X.

$$f_X(x) = \int_0^\infty 0.002e^{-(0.1x+0.02y)} \, dy$$

$$f_X(x) = 0.1e^{-0.1x}$$

$$f_X(x) = \begin{cases} 0.1e^{-0.1x} & x > 0 \\ 0 & \text{otherwise} \end{cases}$$

(d) Determine the marginal PDF of Y.

Let f_Y denote the PDF of Y.

$$f_Y(y) = \int_0^\infty 0.002e^{-(0.1x+0.02y)} \, dx$$

$$f_Y(y) = 0.02e^{-0.02y}$$

$$f_Y(y) = \begin{cases} 0.02e^{-0.02y} & y > 0 \\ 0 & \text{otherwise} \end{cases}$$

(e) Compute the probability that machine 1 fails within two days, but machine 2 survives more than four days.

$$\Pr(X \le 2, Y > 4) = \int_0^2 \int_4^\infty 0.002e^{-(0.1x+0.02y)}\, dy\, dx$$

$$\Pr(X \le 2, Y > 4) = 0.1673$$

(f) Construct an expression for part (e) in terms of only the joint and marginal cumulative distribution functions.

Let F_X denote the CDF of X.

$$\Pr(X \le 2, Y > 4) = F_X(2) - F_{X,Y}(2,4) \quad \square$$

3.1.3 Expectation

As with a single random variable, we can use the probability distribution of two random variables to compute the expected value of a function of the random variables.

Let X and Y be discrete random variables defined on a single random experiment. Let $f_{X,Y}$ denote the joint PMF of X and Y. Let $\{x_1, x_2, \ldots\}$ denote the range of X, and let $\{y_1, y_2, \ldots\}$ denote the range of Y. If g is a real-valued function defined on all (x, y) such that $x \in \{x_1, x_2, \ldots\}$ and $y \in \{y_1, y_2, \ldots\}$, then

$$E\left[g(X,Y)\right] = \sum_{j=1}^{\infty}\sum_{k=1}^{\infty} g(x_j, y_k) f_{X,Y}(x_j, y_k)$$

Let X and Y be continuous random variables defined on a single random experiment. Let $f_{X,Y}$ denote the joint PDF of X and Y. If g is a real-valued function defined on all (x, y) such that x and y are both real numbers, then

$$E\left[g(X,Y)\right] = \int_{-\infty}^{\infty}\int_{-\infty}^{\infty} g(x,y)\; f_{X,Y}(x,y)\; dy\; dx$$

We can use the multivariable Law of the Unconscious Statistician in several meaningful ways. One interesting use corresponds to the case in which the function of the random variables is linear.

If $\{X_1, X_2, \dots, X_n\}$ is a collection of random variables defined on a single random experiment and $\{a_1, a_2, \dots, a_n\}$ is a collection of constants, then

$$E\left(\sum_{j=1}^{n} a_j X_j\right) = \sum_{j=1}^{n} a_j E(X_j) \tag{3.1}$$

Example 3.2 (continued)

What is the expected value of the average time to failure of the two machines?

$$E\left(\frac{X+Y}{2}\right) = E(0.5X + 0.5Y) = 0.5E(X) + 0.5E(Y)$$

$$E(X) = \int_0^\infty 0.1xe^{-0.1x}\, dx$$

$$E(X) = 10 \text{ days}$$

Likewise,

$$E(Y) = \frac{1}{0.02} = 50 \text{ days}$$

$$E\left(\frac{X+Y}{2}\right) = 0.5(10) + 0.5(50) = 30 \text{ days} \ \square$$

We can also use the probability distribution of the two random variables to explore their relationship. Let X and Y be two random variables defined on a single random experiment.

Let $F_{X,Y}$ denote the joint CDF of X and Y, let F_X denote the CDF of X, and let F_Y denote the CDF of Y. The random variables X and Y are said to be **independent** if and only if

$$F_{X,Y}(x, y) = F_X(x)F_Y(y)$$

for all real x and all real y. Otherwise, X and Y are said to be **dependent**.

We can further explore the notion of independent random variables if the two random variables are classified as discrete or continuous.

Let X and Y be discrete random variables defined on a single random experiment. Let $f_{X,Y}$ denote the joint PMF of X and Y, let f_X denote the PMF of X, and let f_Y denote the PMF of Y. Let $\{x_1, x_2, \ldots\}$ denote the range of X, and let $\{y_1, y_2, \ldots\}$ denote the range of Y. The random variables X and Y are **independent** if and only if

$$f_{X,Y}(x,y) = f_X(x) f_Y(y)$$

for all (x, y) such that $x \in \{x_1, x_2, \ldots\}$ and $y \in \{y_1, y_2, \ldots\}$.

Let X and Y be continuous random variables defined on a single random experiment. Let $f_{X,Y}$ denote the joint PDF of X and Y, let f_X denote the PDF of X, and let f_Y denote the PDF of Y. The random variables X and Y are independent if and only if

$$f_{X,Y}(x,y) = f_X(x) f_Y(y)$$

for all real x and all real y.

Example 3.1 (continued)

Are X and Y independent random variables?

$$f_{X,Y}(1,1) = 0.12 \neq f_X(1) f_Y(1) = 0.39(0.42) = 0.1638$$

No, X and Y are not independent. □

Example 3.2 (continued)

Are X and Y independent random variables?

$$f_{X,Y}(x,y) = f_X(x) f_Y(y)$$

Yes, X and Y are independent. □

The concept of independence extends to the multivariable Law of the Unconscious Statistician.

Suppose g and h are real-valued functions defined for all real numbers. If X and Y are two independent random variables defined on a single random experiment, then

$$E\left[g(X)h(Y)\right] = E\left[g(X)\right]E\left[h(Y)\right]$$

This last result facilitates the exploration of a measure of the linear relationship between two random variables.

Let X and Y be two random variables defined on a single random experiment. The **covariance** of X and Y is denoted as $Cov(X, Y)$ and given by

$$Cov(X,Y) = E\left\{\left[X - E(X)\right]\left[Y - E(Y)\right]\right\} = E(XY) - E(X)E(Y) \qquad (3.2)$$

A positive covariance indicates that X tends to increase (decrease) as Y increases (decreases). A negative covariance indicates that X tends to decrease (increase) as Y increases (decreases).

Let X and Y be two random variables defined on a single random experiment. If X and Y are independent, then

$$Cov(X,Y) = 0 \qquad (3.3)$$

Example 3.1 (continued)

Compute the covariance of X and Y.

$$E(XY) = \sum_{x=1}^{3}\sum_{y=1}^{4} x\,y\,f_{X,Y}(x,y) = 3.6$$

$$E(X) = \sum_{x=1}^{3} x\, f_X(x) = 1.83$$

$$E(Y) = \sum_{y=1}^{4} y\, f_Y(y) = 1.96$$

$$Cov(X,Y) = 3.6 - 1.83(1.96) = 0.0132 \; \square$$

Previously, we considered the expected value of the sum of a collection of random variables. The definition of covariance simplifies the evaluation of the variance of a sum of a collection of random variables.

If $\{X_1, X_2, \ldots, X_n\}$ is a collection of random variables defined on a single random experiment, then

$$Var\left(\sum_{j=1}^{n} X_j\right) = \sum_{j=1}^{n} Var\left(X_j\right) + 2\sum_{j=1}^{n-1}\sum_{k=j+1}^{n} Cov\left(X_j, X_k\right) \tag{3.4}$$

If the random variables in the collection are independent, then

$$Var\left(\sum_{j=1}^{n} X_j\right) = \sum_{j=1}^{n} Var\left(X_j\right)$$

An alternative measure of the linear association between two random variables that is more commonly used than covariance is correlation.

Let X and Y be two random variables defined on a single random experiment. The **correlation** between X and Y is denoted by ρ_{XY} and given by

$$\rho_{XY} = \frac{Cov\left(X,Y\right)}{StDev\left(X\right) StDev\left(Y\right)}$$

Correlation and covariance have the same interpretation regarding the relationship between the two variables. However, correlation does not have units and is restricted to the range [−1, 1]. Therefore, the magnitude of correlation provides some idea of the strength of the relationship between the two random variables.

Example 3.1 (continued)

What is the correlation between X and Y?

$$E\left(X^2\right) = \sum_{x=1}^{3} x^2 f_X\left(x\right) = 3.93$$

$$StDev\left(X\right) = \sqrt{Var\left(X\right)} = \sqrt{3.93 - \left(1.83\right)^2} = 0.7623$$

$$E\left(Y^2\right) = \sum_{y=1}^{4} y^2 f_Y\left(y\right) = 4.84$$

$$StDev(X) = \sqrt{Var(X)} = \sqrt{4.84 - (1.96)^2} = 0.9992$$

$$\rho_{XY} = \frac{0.0132}{0.7623(0.9992)} = 0.0173 \square$$

We have already considered the expected value and the variance of the sum of a collection of independent random variables. The next two results can prove valuable in determining the probability distribution of the sum.

Let $\{X_1, X_2, \ldots, X_n\}$ be a collection of independent random variables defined on a single random experiment. Let

$$M_{X_j}(s)$$

denote the moment generating function of X_j, $j = 1, 2, \ldots, n$. If

$$Y = \sum_{j=1}^{n} X_j$$

then the moment generating function of Y is given by

$$M_Y(s) = \prod_{j=1}^{n} M_{X_j}(s) \tag{3.5}$$

Let $\{X_1, X_2, \ldots, X_n\}$ be a collection of independent, non-negative, integer-valued, discrete random variables defined on a single random experiment. Let

$$A_{X_j}(s)$$

denote the probability generating function of X_j, $j = 1, 2, \ldots, n$. If

$$Y = \sum_{j=1}^{n} X_j$$

then the probability generating function of Y is given by

$$A_Y(s) = \prod_{j=1}^{n} A_{X_j}(s) \tag{3.6}$$

3.2 Common Applications of Multiple Random Variables

We now consider two applications of multiple random variables that are well known but not considered further in this text.

3.2.1 The Multinomial Distribution

Consider another scenario related to the random sampling and classification of items from a population.

> Consider a population of items of which each item belongs to one of r classifications. Let p_j denote the proportion of items belonging to classification j, $j = 1, 2, \dots, r$. Note that
>
> $$\sum_{j=1}^{r} p_j = 1$$
>
> Suppose n items are sampled from the population with replacement. Let N_j denote the number of sampled items that belong to classification j, $j = 1, 2, \dots, r$. The range of N_j is $\{0, 1, \dots, n\}$, $j = 1, 2, \dots, n$, such that
>
> $$\sum_{j=1}^{r} N_j = n$$
>
> The collection of random variables $\{N_1, N_2, \dots, N_r\}$ is said to have a **multinomial probability distribution**. The joint PMF of $\{N_1, N_2, \dots, N_r\}$ is given by

$$f(n_1, n_2, \dots, n_r) = \Pr(N_1 = n_1, N_2 = n_2, \dots, N_r = n_r) = \begin{cases} n! \prod_{i=1}^{r} \dfrac{p_i^{n_i}}{n_i!} & \sum_{i=1}^{r} n_i = n \\ 0 & \text{otherwise} \end{cases}$$

> Furthermore, $N_j \sim \text{bin}(n, p_j)$, $j = 1, 2, \dots, r$.

Example 3.3

Each item in a shipment can be classified as good, marginal, or bad. The shipment is such that 90% of items are good, 8% of items are marginal, and 2% of items are bad. Suppose 100 items are sampled from the shipment with replacement. What is the probability that the sample includes one bad item and eleven marginal items?

Let N_1 denote the number of good items in the sample.
Let N_2 denote the number of marginal items in the sample.
Let N_3 denote the number of bad items in the sample.

$$\Pr(N_1 = 88, N_2 = 11, N_3 = 1) = \frac{100!}{88!\,11!\,1!}(0.9)^{88}(0.08)^{11}(0.02)^{1} = 0.0204 \square$$

3.2.2 The Bivariate Normal Distribution

Next, we return to the normal probability distribution.

Let X and Y be continuous random variables, and let $f_{X,Y}$ denote the joint PDF of X and Y. If

$$f_{X,Y}(x,y) = \frac{1}{2\pi\sigma_X\sigma_Y\sqrt{1-\rho^2}} e^{\left\{\frac{-1}{2(1-\rho^2)}\left[\left(\frac{x-\mu_X}{\sigma_X}\right)^2 - \frac{2\rho(x-\mu_X)(y-\mu_Y)}{\sigma_X\sigma_Y} + \left(\frac{y-\mu_Y}{\sigma_Y}\right)^2\right]\right\}}$$

for all real x and all real y where μ_X, μ_Y, σ_X, σ_Y and ρ are constants such that $\sigma_X > 0$, $\sigma_Y > 0$, and $-1 < \rho < 1$, then X and Y are said to have a **bivariate normal probability distribution**. In addition,

$$X \sim N(\mu_X, \sigma_X^2)$$

$$Y \sim N(\mu_Y, \sigma_Y^2)$$

and

$$\rho_{XY} = \rho$$

3.3 Analyzing Discrete Random Variables Using Conditional Probability

In Chapter 1, we computed probabilities associated with events by conditioning on the occurrence of other events. We can do the same for random variables. By conditioning on the value of one random variable, we can often simplify the computation of probabilities associated with or the expected value of another random variable. We begin our exploration of this topic with the case in which both random variables are discrete.

Let X and Y be discrete random variables defined on a single random experiment. Let $f_{X,Y}$ denote the joint PMF of X and Y, and let f_Y denote the PMF of Y. Let $\{x_1, x_2, \ldots\}$ denote the range of X, and let $\{y_1, y_2, \ldots\}$ denote the range of Y. The **conditional probability mass function** (conditional PMF) of X given $Y = y_k$ is given by

$$f_{X|Y}\left(x_j \mid y_k\right) = \Pr\left(X = x_j \middle| Y = y_k\right) = \frac{f_{X,Y}\left(x_j, y_k\right)}{f_Y\left(y_k\right)}$$

$j = 1, 2, \ldots, k = 1, 2, \ldots$. If x is not in the range of X, then

$$f_{X|Y}(x|y) = 0$$

If y is not in the range of Y, then $f_{X|Y}(x|y)$ is undefined. The **conditional cumulative distribution function** (conditional CDF) of X given $Y = y_k$ is given by

$$F_{X|Y}\left(x \mid y_k\right) = \sum_{x_j \le x} f_{X|Y}\left(x_j \mid y_k\right) = \Pr\left(X \le x \mid Y = y_k\right)$$

$k = 1, 2, \ldots$. If y is not in the range of Y, then $F_{X|Y}(x|y)$ is undefined.

The conditional PMF and the conditional CDF can be used to compute probabilities in precisely the same manner as marginal PMFs and CDFs. Although we are able to utilize the conditional PMF and the conditional CDF in many situations, it is quite often simpler to compute the probability of interest using

$$\Pr\left(X \in A \mid Y \in B\right) = \frac{\Pr\left(X \in A, Y \in B\right)}{\Pr\left(Y \in B\right)}$$

where X and Y are random variables and A and B are subsets of the real line such that

$$\Pr\left(Y \in B\right) > 0$$

We can also compute conditional expected values of random variables and functions of random variables.

Let X and Y be discrete random variables defined on a single random experiment. Let $f_{X|Y}$ denote the conditional PMF of X given the value of Y. Let $\{x_1, x_2, \ldots\}$ denote the range of X, and let $\{y_1, y_2, \ldots\}$ denote the range of Y. The **conditional expected value** of X given $Y = y_k$ is given by

$$E(X|Y = y_k) = \sum_{j=1}^{\infty} x_j f_{X|Y}(x_j|y_k)$$

$k = 1, 2, \ldots$

Let X and Y be discrete random variables defined on a single random experiment. Let $f_{X|Y}$ denote the conditional PMF of X given the value of Y. Let $\{x_1, x_2, \ldots\}$ denote the range of X, and let $\{y_1, y_2, \ldots\}$ denote the range of Y. If g is a real-valued function defined for all values in the range of X, then

$$E[g(X)|Y = y_k] = \sum_{j=1}^{\infty} g(x_j) f_{X|Y}(x_j|y_k)$$

$k = 1, 2, \ldots$

Although the conditional PMF may prove to be useful in computing certain conditional expectations, it may be simpler to apply the general rule

$$E[g(X)|Y \in B] = \sum_{j=1}^{\infty} g(x_j) \Pr(X = x_j|Y \in B)$$

where X is a discrete random variable having range $\{x_1, x_2, \ldots\}$, g is a real-valued function defined for all values in the range of X, Y is a random variable, and B is a subset of the real line.

Example 3.1 (continued)

Determine the following.

(a) $\Pr(X = x|Y = 1)$, $x = 1, 2, 3$

$$\Pr(X = 1|Y = 1) = \frac{\Pr(X = 1, Y = 1)}{\Pr(Y = 1)} = \frac{f_{X,Y}(1,1)}{f_Y(1)} = \frac{0.18}{0.42} = \frac{3}{7}$$

$$\Pr(X = 2|Y = 1) = \frac{f_{X,Y}(2,1)}{f_Y(1)} = \frac{0.15}{0.42} = \frac{5}{14}$$

$$\Pr(X = 3|Y = 1) = \frac{f_{X,Y}(3,1)}{f_Y(1)} = \frac{0.09}{0.42} = \frac{3}{14}$$

(b) $\Pr(X = x|Y \geq 3)$, $x = 1, 2, 3$

$$\Pr\left(X=1\middle|Y\geq 3\right)=\frac{\Pr\left(X=1,Y\geq 3\right)}{\Pr\left(Y\geq 3\right)}=\frac{f_{X,Y}\left(1,3\right)+f_{X,Y}\left(1,4\right)}{f_Y\left(3\right)+f_Y\left(4\right)}=\frac{0.06+0.04}{0.18+0.1}=\frac{5}{14}$$

$$\Pr\left(X=2\middle|Y\geq 3\right)=\frac{f_{X,Y}\left(2,3\right)+f_{X,Y}\left(2,4\right)}{f_Y\left(3\right)+f_Y\left(4\right)}=\frac{0.07+0.05}{0.18+0.1}=\frac{3}{7}$$

$$\Pr\left(X=3\middle|Y\geq 3\right)=\frac{f_{X,Y}\left(3,3\right)+f_{X,Y}\left(3,4\right)}{f_Y\left(3\right)+f_Y\left(4\right)}=\frac{0.05+0.01}{0.18+0.1}=\frac{3}{14}$$

(c) $E(X|Y = 1)$

$$E\left(X\middle|Y=1\right)=\sum_{x=1}^{3}x\Pr\left(X=x\middle|Y=1\right)=1\left(\frac{3}{7}\right)+2\left(\frac{5}{14}\right)+3\left(\frac{3}{14}\right)=\frac{25}{14}$$

(d) $Var(X|Y = 1)$

$$E\left(X^2\middle|Y=1\right)=\sum_{x=1}^{3}x^2\Pr\left(X=x\middle|Y=1\right)=1^2\left(\frac{3}{7}\right)+2^2\left(\frac{5}{14}\right)+3^2\left(\frac{3}{14}\right)=\frac{53}{14}$$

$$Var\left(X\middle|Y=1\right)=\frac{53}{14}-\left(\frac{25}{14}\right)^2=\frac{117}{196}$$

(e) $E(Y|Y \geq 3)$

$$\Pr\left(Y=y\middle|Y\geq 3\right)=0\text{ , }y=1,2$$

$$\Pr\left(Y=3\middle|Y\geq 3\right)=\frac{\Pr\left(Y=3,Y\geq 3\right)}{\Pr\left(Y\geq 3\right)}=\frac{\Pr\left(Y=3\right)}{\Pr\left(Y\geq 3\right)}=\frac{f_Y\left(3\right)}{f_Y\left(3\right)+f_Y\left(4\right)}=\frac{0.18}{0.18+0.1}=\frac{9}{14}$$

$$\Pr\left(Y=4\middle|Y\geq 3\right)=\frac{\Pr\left(Y=4,Y\geq 3\right)}{\Pr\left(Y\geq 3\right)}=\frac{\Pr\left(Y=4\right)}{\Pr\left(Y\geq 3\right)}=\frac{f_Y\left(4\right)}{f_Y\left(3\right)+f_Y\left(4\right)}=\frac{0.1}{0.18+0.1}=\frac{5}{14}$$

$$E\left(Y\middle|Y\geq 3\right)=3\left(\frac{9}{14}\right)+4\left(\frac{5}{14}\right)=\frac{47}{14}\ \square$$

Conditional probability also provides an alternative approach to assessing independence of multiple random variables.

> Let X and Y be discrete random variables defined on a single random experiment. Let $f_{X|Y}$ denote the conditional PMF of X given the value of Y, and let f_X denote the PMF of X. Let $\{x_1, x_2, \ldots\}$ denote the range of X, and let $\{y_1, y_2, \ldots\}$ denote the range of Y. The random variables X and Y are independent if and only if
>
> $$f_{X|Y}(x_j \mid y_k) = f_X(x_j)$$
>
> for all $j = 1, 2, \ldots$, and for all $k = 1, 2, \ldots$.

Let X and Y be independent, discrete random variables defined on a single random experiment. Let $\{y_1, y_2, \ldots\}$ denote the range of Y. The independence of X and Y implies that

$$E(X \mid Y = y_k) = E(X)$$

$k = 1, 2, \ldots$, and that

$$E\left[g(X) \mid Y = y_k\right] = E\left[g(X)\right]$$

for any real-valued function g defined for all values in the range of X, $k = 1, 2, \ldots$.

In general, the independence of two random variables X and Y implies that

$$\Pr(X \in A \mid Y \in B) = \Pr(X \in A)$$

where A and B are subsets of the real line. The independence of a discrete random variable X and a random variable Y implies that

$$E\left[g(X) \mid Y \in B\right] = E\left[g(X)\right]$$

where B is a subset of the real line.

3.4 Analyzing Continuous Random Variables Using Conditional Probability

We now consider conditional probability and expectation for the case in which both random variables are continuous.

Let X and Y be continuous random variables defined on a single random experiment. Let $f_{X,Y}$ denote the joint PDF of X and Y, and let f_Y denote the PDF of Y. The **conditional probability density function** (conditional PDF) of X given $Y = y$, for some y such that $f_Y(y) > 0$, is given by

$$f_{X|Y}(x \mid y) = \frac{f_{X,Y}(x, y)}{f_Y(y)}$$

The **conditional cumulative distribution function** (conditional CDF) of X given $Y = y$, for some y such that $f_Y(y) > 0$, is given by

$$F_{X|Y}(x \mid y) = \int_{-\infty}^{x} f_{X|Y}(t \mid y)\, dt = \Pr(X \le x \mid Y = y)$$

The **conditional expected value** of X given $Y = y$, for some y such that $f_Y(y) > 0$, is given by

$$E(X \mid Y = y) = \int_{-\infty}^{\infty} x\, f_{X|Y}(x \mid y)\, dx$$

If g is a real-valued function defined for all real x, then, for some y such that $f_Y(y) > 0$,

$$E\left[g(X) \mid Y = y\right] = \int_{-\infty}^{\infty} g(x)\, f_{X|Y}(x \mid y)\, dx$$

The conditional PDF and the conditional CDF can be used to compute probabilities in precisely the same manner as marginal PDFs and CDFs.

Although we are able to utilize the conditional PDF and the conditional CDF in many situations, it is quite often simpler to compute the probability or expectation of interest using the general approach described in the previous section.

Example 3.4

Let Y denote the time between consecutive inspections of a unit of equipment, and let X denote the hours of operation accumulated by the unit since the last inspection. Both X and Y are measured in hours. The joint PDF of X and Y is given by

$$f_{X,Y}(x,y) = \begin{cases} \dfrac{0.02xe^{-0.01y}}{y^2} & 0 < x < y, y > 0 \\ 0 & \text{otherwise} \end{cases}$$

(a) Determine the conditional PDF of X given Y.

Let f_Y denote the PDF of Y.

$$f_Y(y) = \int_0^y \frac{0.02xe^{-0.01y}}{y^2}\,dx = 0.01e^{-0.01y}$$

$$f_Y(y) = \begin{cases} 0.01e^{-0.01y} & y > 0 \\ 0 & \text{otherwise} \end{cases}$$

Let $f_{X|Y}$ denote the conditional PDF of X given Y.

$$f_{X|Y}(x|y) = \frac{f_{x,y}(x,y)}{f_Y(y)} = \frac{2x}{y^2}$$

$$f_{X|Y}(x|y) = \begin{cases} \dfrac{2x}{y^2} & 0 < x < y \\ 0 & \text{otherwise} \end{cases}$$

(b) Determine $\Pr(X > 8|Y = 10)$.

$$\Pr(X > 8|Y = 10) = \int_8^{10} \frac{2x}{10^2}\,dx = \frac{9}{25}$$

(c) Determine $E(X|Y = 10)$.

$$E(X|Y = 10) = \int_0^{10} \frac{2x^2}{10^2}\,dx = 6.\bar{6}\text{ hr}$$

(d) Determine $\Pr(Y > 50 | Y < 75)$.

$$\Pr(Y > 50 | Y < 75) = \frac{\Pr(Y > 50, Y < 75)}{\Pr(Y < 75)} = \frac{\Pr(50 < Y < 75)}{\Pr(Y < 75)}$$

$$\Pr(Y > 50 | Y < 75) = \frac{\int_{50}^{75} 0.01e^{-0.01y}\, dy}{\int_{0}^{75} 0.01e^{-0.01y}\, dy} = 0.2543$$

(e) Determine $E(Y | Y < 75)$.

Let $F_{Y|Y<75}$ denote the CDF of Y given that $Y < 75$.

Let $f_{Y|Y<75}$ denote the PDF of Y given that $Y < 75$.

$$F_{Y|Y<75}(y) = \Pr(Y \le y | Y < 75) = \frac{\Pr(Y \le y, Y < 75)}{\Pr(Y < 75)} = \frac{\Pr(Y \le y)}{\Pr(Y < 75)}$$

$$F_{Y|Y<75}(y) = \frac{\int_{0}^{y} 0.01e^{-0.01u}\, du}{\int_{0}^{75} 0.01e^{-0.01u}\, du} = \frac{1 - e^{-0.01y}}{1 - e^{-0.75}}$$

$$F_{Y|Y<75}(y) = \begin{cases} 0 & y \le 0 \\ \dfrac{1 - e^{-0.01y}}{1 - e^{-0.75}} & 0 < y < 75 \\ 1 & y \ge 75 \end{cases}$$

$$f_{Y|Y<75}(y) = \frac{dF_{Y|Y<75}(y)}{dy} = \begin{cases} 0 & y \le 0 \\ \dfrac{0.01e^{-0.01y}}{1 - e^{-0.75}} & 0 < y < 75 \\ 0 & y \ge 75 \end{cases}$$

$$E(Y | Y < 75) = \int_{0}^{75} \frac{0.01ye^{-0.01y}}{1 - e^{-0.75}}\, dy$$

$$E(Y|Y<75) = 32.86 \text{ hr } \square$$

Conditional probability again provides an alternative approach to assessing independence of multiple random variables.

Let X and Y be continuous random variables defined on a single random experiment. Let $f_{X|Y}$ denote the conditional PDF of X given the value of Y, let f_X denote the PDF of X, and let f_Y denote the PDF of Y. The random variables X and Y are independent if and only if

$$f_{X|Y}(x|y) = f_X(x)$$

for all y such that

$$f_Y(y) > 0$$

3.5 Computing Expectations by Conditioning

In the previous sections of this chapter, we describe how we can use knowledge about one random variable to construct the conditional probability distribution for another random variable. We then demonstrate how to obtain the conditional expected value of the random variable. However, in computing expectations, the most common objective is to obtain the unconditional expected value of a random variable when conditional expectations are known. In this section, we consider the problem associated with this objective.

Let X and Y be random variables defined on a single random experiment. If Y is a discrete random variable having PMF f_Y and range $\{y_1, y_2, \ldots\}$, then

$$E(X) = \sum_{j=1}^{\infty} E(X|Y=y_j) f_Y(y_j)$$

If Y is a continuous random variable having PDF f_Y, then

$$E(X) = \int_{-\infty}^{\infty} E(X|Y=y) f_Y(y)\, dy$$

This method of obtaining the expected value of a random variable is sometimes referred to as **smoothing**. Smoothing can also be applied to a function of a random variable.

Let X and Y be random variables defined on a single random experiment, and let g be a real-valued function defined for all values in the range of X. If Y is a discrete random variable having PMF f_Y and range $\{y_1, y_2, \ldots\}$, then

$$E\left[g(X)\right]=\sum_{j=1}^{\infty}E\left[g(X)\mid Y=y_j\right]f_Y(y_j)$$

If Y is a continuous random variable having PDF f_Y, then

$$E\left[g(X)\right]=\int_{-\infty}^{\infty}E\left[g(X)\mid Y=y\right]f_Y(y)\,dy$$

Example 3.5

Let X denote the single-day revenue of a store that sells custom home theater systems. The probability distribution of X is dependent on Y, the number of serious customers entering the store on the day of interest. Let f_Y denote the PMF of Y where

$f_Y(0) = 0.1$ $f_Y(1) = 0.4$ $f_Y(2) = 0.3$ $f_Y(3) = 0.2$

Furthermore, suppose

$$E(X|Y=0)=\$0$$

$$E(X|Y=1)=\$1400$$

$$E(X|Y=2)=\$2300$$

$$E(X|Y=3)=\$3100$$

and

$$StDev(X|Y=0)=\$0$$

$$StDev(X|Y=1)=\$300$$

$$StDev(X|Y=2)=\$800$$

$$StDev(X|Y=3) = \$1400$$

(a) Determine $E(X)$.

$$E(X) = \sum_{y=0}^{3} E(X|Y=y)\Pr(Y=y) = \$0(0.1) +$$

$$\$1400(0.4) + \$2300(0.3) + \$3100(0.2)$$

$$E(X) = \$1870$$

(b) Determine $StDev(X)$.

Note that

$$Var(X|Y=y) = (StDev(X|Y=y))^2 = E(X^2|Y=y) - (E(X|Y=y))^2$$

$$E(X^2|Y=y) = (StDev(X|Y=y))^2 + (E(X|Y=y))^2$$

Therefore,

$$E(X^2|Y=0) = (0)^2 + (0)^2 = 0$$

$$E(X^2|Y=1) = (300)^2 + (1400)^2 = 2050000$$

$$E(X^2|Y=2) = (800)^2 + (2300)^2 = 5930000$$

$$E(X^2|Y=3) = (1400)^2 + (3100)^2 = 11570000$$

$$E(X^2) = 0(0.1) + 2050000(0.4) + 5930000(0.3) + 11570000(0.2) = 4913000$$

$$StDev(X) = \sqrt{Var(X)} = \sqrt{4913000 - (1870)^2} = \$1190 \ \square$$

Example 3.5 demonstrates that smoothing cannot be applied directly to the variance or standard deviation.

Example 3.6

Let X denote the time to failure of a machine, and let Y denote the repair time associated with this failure. Both X and Y are measured in hours. Let f_X denote the PDF of X where

$$f_X(x) = \begin{cases} 0.04e^{-0.04x} & x > 0 \\ 0 & \text{otherwise} \end{cases}$$

Suppose Y is uniformly distributed over the interval $(0.5X, X)$. Determine $E(Y)$.

$$E(Y \mid X = x) = \frac{0.5x + x}{2} = 0.75x$$

$$E(Y) = \int_0^\infty (0.75x)\left(0.04e^{-0.04x}\right) dx$$

$$E(Y) = 18.75hr \quad \square$$

Conditional expectation also facilitates the analysis of what are known as compound random variables.

> Let $\{X_1, X_2, \ldots, X_N\}$ be a collection of independent and identically distributed random variables defined on a single random experiment where N is a non-negative, discrete random variable. If
>
> $$Y = \sum_{j=1}^{N} X_j$$
>
> then Y is referred to as a **compound random variable** and the probability distribution of Y is referred to as a **compound probability distribution.**

One of the most famous results in probability theory involves the expected value of a compound random variable. This result is known as **Wald's Lemma** (Wald, 1944).

> Let $\{X_1, X_2, \ldots, X_N\}$ be a collection of independent and identically distributed random variables defined on a single random experiment where N is a non-negative, discrete random variable. If
>
> $$Y = \sum_{j=1}^{N} X_j$$

then

$$E(Y) = E(N)E(X_j)$$

and

$$Var(Y) = E(N)Var(X_j) + \left[E(X_j)\right]^2 Var(N)$$

Example 3.7

Suppose the number of customers entering a store on a given day is a non-negative, discrete random variable having a mean of 10 and a variance of 10. The amount of money spent by a single customer is independent of the amount spent by any other customer and is uniformly distributed over the interval ($0, $100). Find the mean and standard deviation of the amount of money taken in by the store on a given day.

Let N denote the number of customers entering the store.

Let X_j denote the amount of money spent by customer j, $j = 1, 2, \ldots, N$.

Note that

$$E(N) = 10$$

$$Var(N) = 10$$

$$E(X_j) = \frac{0 + 100}{2} = 50$$

$$Var(X_j) = \frac{(100 - 0)^2}{12} = 833.\overline{3}$$

Therefore,

$$E(Y) = 10(50) = \$500$$

$$StDev(Y) = \sqrt{Var(Y)} = \sqrt{10(833.\overline{3}) + (50)^2(10)} = \$182.57 \ \square$$

3.6 Computing Probabilities by Conditioning

In the previous section, we considered methods of computing the expectation of a random variable or a function of a random variable by conditioning on the value of some other random variable. In this section, we consider the problem of computing probabilities by conditioning on the value of some random variable.

Let A be an event, and let Y be a random variable defined on single random experiment. If Y is a discrete random variable having PMF f_Y and range $\{y_1, y_2, \ldots\}$, then

$$\Pr(A) = \sum_{j=1}^{\infty} \Pr(A \mid Y = y_j) f_Y(y_j)$$

If Y is a continuous random variable having PDF f_Y, then

$$\Pr(A) = \int_{-\infty}^{\infty} \Pr(A \mid Y = y) f_Y(y)\, dy$$

Note that the event of interest could refer to a random variable taking on some specified value(s).

Example 3.8

Let X denote the number of orders received by a warehouse during a one-week period. The inventory management policies utilized by the warehouse are such that 3% of orders cannot be filled, and each order is independent of every other order. Let Y denote the number of unfilled orders in one week. The PMF of X is given by

$$f_X(x) = \begin{cases} \dfrac{e^{-12}12^x}{x!} & x = 0, 1, \ldots \\ 0 & \text{otherwise} \end{cases}$$

Determine the probability that there are no unfilled orders in one week.

$$\Pr(Y = 0) = \sum_{x=0}^{\infty} \Pr(Y = 0 \mid X = x) \Pr(X = x) = \sum_{x=0}^{\infty} (0.97)^x \frac{e^{-12}12^x}{x!} = e^{-12} \sum_{x=0}^{\infty} \frac{11.64^x}{x!}$$

$$\Pr(Y = 0) = e^{-12} e^{11.64} = e^{-0.36} = 0.6977 \;\square$$

Example 3.6 (continued)

Determine the probability that the repair time is less than 10 hours.

$$\Pr(Y<10)=\int_0^{\infty}\Pr(Y<10|X=x)f_X(x)\,dx$$

$$\Pr(Y<10)=\int_0^{10}\Pr(Y<10|X=x)f_X(x)\,dx+\int_{10}^{20}\Pr(Y<10|X=x)f_X(x)\,dx$$

$$+\int_{20}^{\infty}\Pr(Y<10|X=x)f_X(x)\,dx$$

$$\Pr(Y<10)=\int_0^{10}1f_X(x)\,dx+\int_{10}^{20}\left(\int_{0.5x}^{10}\frac{1}{x-0.5x}\,dy\right)f_X(x)\,dx+\int_{20}^{\infty}0f_X(x)\,dx$$

$$\Pr(Y<10)=\int_0^{10}0.04e^{-0.04x}\,dx+\int_{10}^{20}\left(\frac{10-0.5x}{x-0.5x}\right)0.04e^{-0.04x}\,dx$$

$$\Pr(Y<10)=0.4221\ \square$$

Previously, we described how to compute the expected value and variance for the sum of two random variables. Suppose, however, that we need to know the distribution of the sum of two random variables.

Let X and Y be random variables defined on a single random experiment. Let F_X denote the CDF of X, and let F_Y denote the CDF of Y. Let

$$Z=X+Y$$

If F_Z denotes the CDF of Z, then $F_Z(z)$ is referred to as the **convolution** of F_X and F_Y. If X and Y are discrete random variables such that $\{x_1, x_2, \ldots\}$ denotes the range of X, $\{y_1, y_2, \ldots\}$ denotes the range of Y, f_X denotes the PMF of X, and f_Y denotes the PMF of Y, then

$$F_Z(z)=\sum_{k=1}^{\infty}\sum_{x_j\le z-y_k}f_X(x_j)f_Y(y_k)$$

and

$$f_Z(z)=\sum_{k=1}^{\infty}f_X(z-y_k)f_Y(y_k)$$

where f_Z denotes the PMF of Z. If X and Y are continuous random variables such that f_X denotes the PDF of X and f_Y denotes the PDF of Y, then

$$F_Z(z) = \int_{-\infty}^{\infty} F_X(z-y) f_Y(y)\, dy$$

and

$$f_Z(z) = \int_{-\infty}^{\infty} f_X(z-y) f_Y(y)\, dy$$

where f_Z denotes the PDF of Z.

Unfortunately, there are not many situations in which this approach provides a unique avenue to determining the probability distribution of the sum of two random variables.

Homework Problems

3.1 Two Random Variables

(1) A gambling game operates as follows. A player pays \$5 to play. A fair coin is then flipped. If the coin lands heads, then a fair die is rolled. If the coin lands tails, then a loaded die is rolled. The loaded die is such that

$$\Pr(1) = \Pr(2) = 0.3$$

and

$$\Pr(3) = \Pr(4) = \Pr(5) = \Pr(6) = 0.1$$

The player is then paid \$2 multiplied by the value appearing on the die. Let

$$X = \begin{cases} 0 & \text{if the coin lands heads} \\ 1 & \text{if the coin lands tails} \end{cases}$$

Let Y denote the value appearing on the die roll.

(a) Construct the joint PMF of X and Y.
(b) Determine the marginal PMFs of X and Y.
(c) What is the expected value of the player's winnings?
(d) What is the covariance of X and Y?
(e) What is the correlation between X and Y?

(2) A company provides equipment maintenance services to a variety of organizations. The timeliness of service depends on the general availability of spare parts. Let Y denote the unit-less measure of spare parts availability at the time a service call is received where $Y \in (0, 1)$. A larger value of Y implies greater spare part availability. Let X denote the time required to complete service where X is measured in hours. The joint PDF of X and Y is given by

$$f_{X,Y}(x,y) = \begin{cases} 2y^2e^{-xy} & x > 0,\ y > 1 \\ 0 & \text{otherwise} \end{cases}$$

(a) Construct the joint CDF of X and Y.
(b) Determine the marginal PDFs of X and Y.
(c) What is the covariance of X and Y?

(3) Prove (3.1) for the case in which $n = 2$ and the random variables are continuous.

(4) Prove that the second equality in (3.2) holds for the case in which X is a discrete random variable having range $\{x_1, x_2, \ldots\}$ and Y is a discrete random variable having range $\{y_1, y_2, \ldots\}$.

(5) Prove (3.4) for the case in which $n = 3$.

(6) Prove (3.5).

(7) Prove (3.6).

(8) An instructor is waiting on n students to complete an assignment. Let X_i denote the time required for student i to complete the task, $i = 1, 2, \ldots, n$. Suppose $\{X_1, X_2, \ldots, X_n\}$ is a collection of independent and identically distributed continuous random variables such that the PDF of each random variable in the collection is given by

$$f(x) = \begin{cases} \lambda e^{-\lambda x} & x > 0 \\ 0 & \text{otherwise} \end{cases}$$

where λ is a constant such that $\lambda > 0$. Let X_{min} denote the time that the first completion of the assignment occurs, and let X_{max} denote the time that the last completion occurs.

(a) Construct the CDF of X_{max}.
(b) Determine the expected value of X_{max}.
(c) Construct the CDF of X_{min}.
(d) Determine the expected value of X_{min}.

3.2 Common Applications of Multiple Random Variables

(1) Historical data suggests that a scientific journal accepts 10% of the manuscripts it receives, rejects 50% of the manuscripts it receives, and requests "revise and resubmit" for the remaining manuscripts. In a given month, the journal receives 75 manuscripts.

(a) What is the probability that 5 are accepted and 40 are rejected?
(b) What is the probability that 5 are accepted?

3.3 Analyzing Discrete Random Variables Using Conditional Probability

(1) An automated machine manufactures two products of the same type simultaneously. Each product is produced in one of two chambers (chamber 1 and chamber 2) within the machine, and each is subject to three potential defects. Let X denote the number of defects in the product produced in chamber 1, and let Y denote the number of defects in the product produced in chamber 2. The joint PMF of X and Y is given by

$f_{X,Y}(0,0) = 0.85$	$f_{X,Y}(1,0) = 0.03$	$f_{X,Y}(2,0) = 0.008$	$f_{X,Y}(3,0) = 0.003$
$f_{X,Y}(0,1) = 0.05$	$f_{X,Y}(1,1) = 0.015$	$f_{X,Y}(2,1) = 0.004$	$f_{X,Y}(3,1) = 0.001$
$f_{X,Y}(0,2) = 0.02$	$f_{X,Y}(1,2) = 0.005$	$f_{X,Y}(2,2) = 0.002$	$f_{X,Y}(3.2) = 0.0008$
$f_{X,Y}(0,3) = 0.007$	$f_{X,Y}(1,3) = 0.003$	$f_{X,Y}(2,3) = 0.001$	$f_{X,Y}(3,3) = 0.0002$

(a) Determine $E(X|Y=1)$.
(b) Determine $StDev(X|Y=1)$.

(c) Determine $E(X|Y>1)$.

(d) Determine $StDev(X|X<2)$.

(2) A supplier has a contract with three companies to provide them with a specialty item. Each Monday, the supplier receives orders for the item from a subset of the companies. Let X denote the number of companies that place an order, and let Y denote the total number of items ordered. The joint PMF of X and Y is given by

$$f_{X,Y}(x,y)=\begin{cases}\dfrac{0.25e^{-5x}(5x)^y}{y!} & x=0,1,2,3;\ y=0,1,\ldots\\ 0 & \text{otherwise}\end{cases}$$

(a) Determine the conditional PMF of Y given X = x.

(b) Determine $E(Y|X=2)$.

(c) Determine $StDev(Y|X=2)$.

(d) Determine $E(Y|X\geq 2)$.

(e) Determine $StDev(X|X<2)$.

3.4 Analyzing Continuous Random Variables Using Conditional Probability

(1) Let X denote the time to failure of a mechanical component, and let Y denote the time required to repair the component. Both X and Y are measured in hours. The joint PDF of X and Y is given by

$$f_{X,Y}(x,y)=\begin{cases}\dfrac{0.02}{x}\exp\left(-0.002x-\dfrac{10y}{x}\right) & x>0,\ y>1\\ 0 & \text{otherwise}\end{cases}$$

(a) Determine the conditional PDF of Y given X.

(b) Determine $E(Y|X=400)$.

(c) Determine $StDev(Y|X=400)$.

(d) Determine $E(X|X>700)$.

(2) Let X denote the amount (measured in tons) of recyclable material in a truck, and let Y denote the monetary value (measured in thousands of U.S. dollars) of this material. The joint probability density function of X and Y is given by

$$f_{X,Y}(x,y) = \begin{cases} \frac{25y}{4x^2} & 6 < x < 10,\ 0 < y < 0.2x \\ 0 & \text{otherwise} \end{cases}$$

(a) Determine the conditional PDF of Y given $X = x$.

(b) Determine $E(Y|X = 8)$.

(c) Determine $StDev(Y|X = 8)$.

(d) Determine $E(X|X > 7.5)$.

3.5 Computing Expectations by Conditioning

(1) The useful life of a medical device depends on its initial quality. Let X denote the useful life of the device (measured in weeks), and let Y denote the initial quality of the device. The device is such that its initial quality is in one of three classifications. These classifications are denoted by 1, 2, and 3 where 1 is best and 3 is worst. Let f_Y denote the PMF of Y where

$f_Y(1) = 0.9$ $f_Y(2) = 0.07$ $f_Y(3) = 0.03$

Furthermore, suppose

$$E(X|Y = 1) = 250$$

$$E(X|Y = 2) = 120$$

$$E(X|Y = 3) = 30$$

and

$$StDev(X|Y = 1) = 30$$

$$StDev(X|Y = 2) = 40$$

$$StDev(X|Y=3)=20$$

(a) Determine $E(X)$.

(b) Determine $StDev(X)$.

(2) Let Y denote the number of claims filed with an insurance agency during a one-month period, and let X denote the number of these claims that are ultimately paid in full. Let f_Y denote the PMF of Y where

$$f_Y(y)=\left\{\frac{e^{-100}100^y}{y!}\right.$$

$y = 0, 1, \ldots$. Furthermore, suppose that X is uniformly distributed over the integer set $\{0, 1, \ldots, Y\}$.

(a) Determine $E(X)$.

(b) Determine $StDev(X)$.

(3) Let X denote the time (measured in minutes) required for a service technician to complete a task. The probability distribution of X is dependent on Y, the years of training possessed by the technician selected to perform the task. Suppose that Y is uniformly distributed over the interval (1, 5) and suppose that the conditional PDF of X given Y is given by

$$f_{Y|X}(y|x)=\begin{cases} ye^{-yx} & x>0 \\ 0 & \text{otherwise} \end{cases}$$

Determine $E(X)$.

(4) A message being sent over a communications network is assigned by a router to one of three paths (path 1, path 2, path 3). The nature of the network is such that 50% of all messages are routed to path 1, 30% are routed to path 2, and 20% are routed to path 3. If routed to path 1, then the message has a 75% chance of reaching its destination immediately. Otherwise, the message experiences a five-second delay and returns to the router. If routed to path 2, then the message has a 60% chance of reaching its destination immediately. Otherwise, the message experiences a ten-second delay and returns to the router. If routed to path 3, then the message has a 40% chance of reaching

its destination immediately. Otherwise, the message experiences a twenty-second delay and returns to the router. Note that the router cannot distinguish between new messages and messages that have returned from an unsuccessful attempt. Let X denote the time until the message reaches its destination. Determine $E(X)$ and $StDev(X)$.

(5) Suppose the number of traffic accidents in a given city during a one-month period is a non-negative, discrete random variable having a mean of 50 and a variance of 100. The number of fatalities in a single traffic accident is independent of the number of fatalities in any other accident and has a mean of 0.2 and a variance of 1. Find the mean and standard deviation of the number of traffic fatalities in the city in one month.

(6) Suppose the number of tornadoes in a given region during a one-year period is a non-negative, discrete random variable having a mean of 8 and a standard deviation of 2. The economic impact of a single tornado is independent of the impact of any other tornado and has a mean of \$2 million and a standard deviation of \$1.5 million. Find the mean and standard deviation of the economic impact of tornadoes in the region in one year.

3.6 Computing Probabilities by Conditioning

(1) Let Y denote the number of times during a one-month period that a machine is shut down for maintenance. Each time a shutdown occurs, there is a 15% chance that the maintenance technician causes the machine to be out of calibration. Note that a shutdown is independent of all other shutdowns. Let X denote the number of times recalibration is required during a one-month period. The PMF of Y is given by

$$f_{Y|X}(y) = \begin{cases} 0.4(0.6)^y & y = 0, 1, \ldots \\ 0 & \text{otherwise} \end{cases}$$

Determine $\Pr(X = 0)$.

(2) Consider Example 3.8. Suppose the conditional PMF of Y given X is

$$f_Y(y|x) = \begin{cases} \dfrac{x!}{(x-y)!\,y!}(0.03)^y(0.97)^{x-y} & y > 0, 1, \ldots, x \\ 0 & \text{otherwise} \end{cases}$$

Determine $\Pr(Y = 3)$.

(3) Let Y denote the concentration of imperfections in a batch of raw material used in a continuous production process. The PDF of Y is given by

$$f_Y(y) = \begin{cases} \lambda e^{-\lambda y} & y > 0 \\ 0 & \text{otherwise} \end{cases}$$

where $\lambda > 0$ is a constant. Let X denote the number of machine breakdowns occurring while processing the batch. The conditional PMF of X given Y is

$$f_{X|Y}(x|y) = \begin{cases} \dfrac{e^{-\alpha y}(\alpha y)^x}{x!} & x = 0, 1, \ldots \\ 0 & \text{otherwise} \end{cases}$$

where $\alpha > 0$ is a constant. Prove that

$$\Pr(X = x) = \frac{\lambda \alpha^x}{(\alpha + \lambda)^{x-1}}$$

for all $x = 0, 1, \ldots$.

Application: Bivariate Warranty Modeling

In traditional models of equipment failure, reliability is typically determined using a time to failure probability distribution. However, the reliability of many types of equipment (e.g., automobiles) may be more accurately modeled by considering both time to failure and use to failure. In the case of an automobile, we may choose to model reliability using time and mileage to failure.

Let T denote the time to failure for a product, and let U denote the use to failure for the product. Let $f_{T,U}(t,u)$ denote the joint PDF of T and U where

$$f_{T,U}(t,u) = \begin{cases} \lambda \eta e^{-(\lambda t + \eta u)} \left\{1 + \rho\left(1 - 2e^{-\lambda t} - 2e^{-\eta u} + 4e^{-(\lambda t + \eta u)}\right)\right\} & t > 0, u > 0 \\ 0 & \text{otherwise} \end{cases}$$

Under such a model, T and U are said to have a bivariate exponential distribution.

Task 1: Construct an expression for the joint CDF of T and U.

Task 2: What kind of random variables are T and U? Support your answer.

Task 3: Are T and U independent? Support your answer. If not, what is the correlation between T and U?

When reliability is determined using a time to failure distribution, reliability is defined as the probability that the product survives more than t time units. The resulting reliability function is

$$R(t) = 1 - F_T(t)$$

where $F_T(t)$ denotes the CDF of T.

Task 4: Define and interpret an appropriate bivariate reliability function.

Suppose the sale of the product includes the following warranty:

> If the product fails within t_1 time units and u_1 units of use, the customer will be refunded the purchase price, s.

Not all customers take advantage of the warranty. Let a denote the probability that an eligible customer uses the warranty, and let c_a denote the cost of processing a warranty claim.

Task 5: Let $W(T,U,t_1,u_1)$ denote the warranty cost for a product that fails at time T and use U under a warranty policy (t_1,u_1). Construct an expression for $E[W(T,U,t_1,u_1)]$ in terms of the joint cumulative distribution function of T and U.

The failure of a product also implies some lost goodwill on the part of the customer. Let $G(T,U,t_1,u_1)$ denote the cost of lost goodwill if failure occurs at time T and use U under a warranty policy (t_1,u_1) where

$$G(T,U,t_1,u_1)=\begin{cases} g_{00} & 0\le t\le t_1\ ,0\le u\le u_1 \\ g_{01} & 0\le t\le t_1\ ,u_1\le u\le u_2 \\ g_{10} & t_1\le t\le t_2\ ,0\le u\le u_1 \\ g_{11} & t_1\le t\le t_2\ ,u_1\le u\le u_2 \\ 0 & \text{otherwise} \end{cases}$$

t_2 is an upper bound on the amount of time a customer intends to own the product, and u_2 is a similar upper bound on use.

Task 6: Construct an expression for $E[G(T,U,t_1,u_1)]$ in terms of the joint CDF of T and U.

Task 7: Let $P(T,U,t_1,u_1)$ denote the profit generated by a product that fails at time T and use U under a warranty policy (t_1,u_1). Construct an expression for $E[P(T,U,t_1,u_1)]$ in terms of $E[W(T,U,t_1,u_1)]$ and $E[G(T,U,t_1,u_1)]$.

The choices for the values of t_1 and u_1 impact the sales for a product (a better warranty leads to increased sales). Let $d(t_1,u_1)$ denote the demand for a product resulting from a warranty policy (t_1,u_1) where

$$d(t_1,u_1)=d_0+(d_1-d_0)\left(1-e^{-kt_1u_1}\right)$$

d_0 denotes the minimum demand, d_1 denotes the maximum demand, and $k > 0$ is a parameter.

Task 8: Let $B(t_1,u_1)$ denote the total profit resulting from all sales if a warranty policy (t_1,u_1) is implemented. Construct an expression for $E[B(t_1,u_1)]$ in terms of $d(t_1,u_1)$ and $E[P(T,U,t_1,u_1)]$.

Suppose the parameter values for a particular product are:

$\lambda = 0.01$ weeks, $\eta = 0.02$ cycles, $\rho = 0.2$, $s = \$100$, $a = 0.7$, $c_a = \$15$

$g_{00} = \$5$, $g_{10} = \$15$, $g_{01} = \$10$, $g_{11} = \$25$, $t_2 = 300$ weeks, $u_2 = 200$ cycles

$d_0 = 15{,}000$ units, $d_1 = 20{,}000$ units, $k = 0.001$

Task 9: Compute the 75-week, 40-cycle reliability for the product.

Task 10: Recommend a warranty policy to the manufacturer. Support this recommendation with numerical results.

4

Introduction to Stochastic Processes

Even though random variables can provide important descriptors of the behavior of industrial systems, random variables do not capture the evolution of unpredictable behavior over time. To model the time evolution of a random variable, we utilize stochastic processes.

4.1 Introduction to Stochastic Processes

Stochastic processes are used to model the behavior of a random variable over time.

> A collection of random variables $\{X(t), t \in T\}$ is called a **stochastic process**.

For all $t \in T$, $X(t)$ is a random variable. Although it does not have to be, the index t is most often some measure of elapsed time. We accept this interpretation.

> If $\{X(t), t \in T\}$ is a stochastic process, then $X(t)$ is the **state** of the process at time t for all $t \in T$.

The following are examples of stochastic processes.

1. Let $X(t)$ denote the number of people in a building at time t.
2. Let $X(t)$ denote the official temperature at an airport at time t.
3. Let $X(t)$ denote the inventory level of product 12 in warehouse B at the end of day t.
4. Let $X(t)$ denote the concentration of a certain pollutant in a river at 8:00 a.m. on day t.

Stochastic processes can be classified according to the manner in which time evolves and the range of the random variables within the stochastic process.

> If $\{X(t), t \in T\}$ is a stochastic process, then the set T is called the **index set** of the process. If T is a countable set, then $\{X(t), t \in T\}$ is said to be a **discrete-time stochastic process**. If T is an uncountable set, then $\{X(t), t \in T\}$ is said to be a **continuous-time stochastic process**.

If $\{X(t), t \in T\}$ is a stochastic process, then the set of all possible states of the stochastic process is called the **state space.** If the state space is countable, then $\{X(t), t \in T\}$ is said to be a **discrete-valued stochastic process.** If the state space is uncountable, then $\{X(t), t \in T\}$ is said to be a **continuous-valued stochastic process.**

For the examples enumerated above, example 1 is continuous-time, discrete-valued; example 2 is continuous-time, continuous-valued; example 3 is discrete-time, discrete-valued; and example 4 is discrete-time, continuous-valued. We consider all four types of stochastic processes in the remainder of this chapter, but our discussions begin with two general classes of stochastic processes.

4.2 Introduction to Counting Processes

Suppose we are interested in monitoring and counting the occurrences of some event over an interval of time.

Assume some event of interest can occur at any point in time. If $N(t)$ denotes the number of occurrences of the event during the time interval $[0,t]$, then the continuous-time stochastic process $\{N(t), t \geq 0\}$ is said to be a **counting process.**

Alternately, for an event that can occur only at discrete points in time which are denoted by $t = 0, 1, \dots$, $N(t)$ denotes the cumulative number of occurrences of the event at times $0, 1, \dots, t$, and the discrete-time stochastic process $\{N(t), t = 1, 2, \dots\}$ is said to be a **counting process.**

In formally defining a counting process, the reference time $t = 0$ and the measure of time (minutes, days, etc.) need to be specified or clearly implied. Examples of counting processes include:

1. $\{N(t), t \geq 0\}$ where $N(t)$ denotes the number of earthquakes in California by time t where t is measured in years
2. $\{N(t), t \geq 0\}$ where $N(t)$ denotes the number of births at a hospital by time t where t is measured in days
3. $\{N(t), t \geq 0\}$ where $N(t)$ denotes the number of hamburgers sold at a restaurant by time t where t is measured in hours
4. $\{N(t), t \geq 0\}$ where $N(t)$ denotes the number of machine failures in a facility by time t where t is measured in weeks
5. $\{N(t), t \geq 0\}$ where $N(t)$ denotes the number of lost sales due to stock-outs in an inventory system by time t where t is measured in days

6. $\{N(t), t = 1, 2, \dots\}$ where $N(t)$ denotes the number of defective items found in the first t items inspected
7. $\{N(t), t = 1, 2, \dots\}$ where $N(t)$ denotes the number of "yes" votes in the first t people polled

There are two clear implications of the definition of a counting process.

> If $\{N(t), t \in T\}$ is a counting process, then at any point in time $N(t) \in \{0, 1, \dots\}$. Furthermore, if $s < t$, then $N(s) \leq N(t)$.

The following examples are not counting processes because they could decrease over time.

1. $\{N(t), t \geq 0\}$ where $N(t)$ denotes the population of Arkansas at time t
2. $\{N(t), t \geq 0\}$ where $N(t)$ denotes the number of customers in a restaurant at time t
3. $\{N(t), t \geq 0\}$ where $N(t)$ denotes the number of machines operating in a facility at time t
4. $\{N(t), t \geq 0\}$ where $N(t)$ denotes the number of customers waiting in a line at time t

A counting process may also possess one or both of two noteworthy characteristics.

> Consider a counting process $\{N(t), t \in T\}$, and suppose $0 \leq t_1 < t_2 \leq t_3 < t_4$. If $N(t_2) - N(t_1)$ and $N(t_4) - N(t_3)$ are independent random variables, then $\{N(t), t \in T\}$ is said to have **independent increments.**

> Consider a counting process $\{N(t), t \in T\}$, and suppose $s \geq 0$ and $x > 0$. If the probability distribution on $N(s + x) - N(s)$ is independent of s, then $\{N(t), t \in T\}$ is said to have **stationary increments.**

The term independent increments implies that the number of events in some interval of time is independent of the number of events occurring in any other nonoverlapping interval of time. The term stationary increments implies that the number of events occurring in some interval of time depends only on the length of the interval and not on the beginning time of the interval.

4.3 Introduction to Renewal Processes

A renewal process is a continuous-time, counting process for which the times between events are independent and identically distributed random variables. The term "renewal" is derived from the fact that after each event, the process is restarted in that it is probabilistically identical in behavior to its realization at time $t = 0$ (It is "like new" or renewed).

> Let $\{N(t), t \geq 0\}$ be a continuous-time, counting process, let $X(1)$ denote the time until the first event, and let $X(n)$ denote the time between the $(n-1)^{\text{th}}$ and n^{th} event, $n = 2, 3, \ldots$. If $\{X(1), X(2), \ldots\}$ is a collection of independent and identically distributed random variables , then $\{N(t), t \geq 0\}$ is a **renewal process**. If
>
> $$S(n) = \sum_{j=1}^{n} X(j)$$
>
> then $S(n)$ denotes the nth **renewal time** (or the time of the nth renewal). We also refer to $X(n)$ as the nth **inter-renewal time.**

For a renewal process $\{N(t), t \geq 0\}$, we can derive general expressions for the PMF of $N(t)$.

> Let $\{N(t), t \geq 0\}$ be a renewal process, let $\{X(1), X(2), \ldots\}$ denote the corresponding sequence of inter-renewal times, and let $\{S(1), S(2), \ldots\}$ denote the corresponding sequence of renewal times. If F_X denotes the CDF of $X(n)$, $n = 1, 2, \ldots$, and $F_X^{(n)}$ denotes the CDF of $S(n)$, $n = 1, 2, \ldots$, then
>
> $$\Pr\left[N(t) = 0\right] = 1 - F_X(t) \tag{4.1}$$
>
> and
>
> $$\Pr\left[N(t) = n\right] = F_X^{(n)}(t) - F_X^{(n+1)}(t) \tag{4.2}$$
>
> $n = 1, 2, \ldots$. Note that $F_X^{(n)}$ is the n-fold convolution of F_X.

Equation (4.2) is referred to as the **renewal equation**. The usefulness of the renewal equation is limited to those cases in which the n-fold convolution is easy to obtain. We consider the most common of these cases, the Poisson process, later in this text.

For renewal processes, there is a fundamental duality relationship between the number of events that occur during an interval of time and the time that elapses until a fixed number of events have occurred. This time-frequency relationship is expressed as

$$\Pr[N(t) \geq n] = \Pr[S(n) \leq t]$$

This equation indicates that the number of events by time t can only exceed n if the nth event occurs prior to time t. While this condition may seem obvious, its realization in terms of probability directly leads to equation (4.2)

For a renewal process $\{N(t), t \geq 0\}$, we can also construct a general expression for the expected value of $N(t)$.

Let $\{N(t), t \geq 0\}$ be a renewal process, let $\{X(1), X(2), \ldots\}$ denote the corresponding sequence of inter-renewal times, and let $\{S(1), S(2), \ldots\}$ denote the corresponding sequence of renewal times. If f_X denotes the PDF of $X(n), n = 1, 2, \ldots, F_X$ denotes the CDF of $X(n), n = 1, 2, \ldots$, and $F_X^{(n)}$ denotes the CDF of $S(n), n = 1, 2, \ldots$, then

$$m(t) = E\left[N(t)\right] = \sum_{n=1}^{\infty} F_X^{(n)}(t) = F_X(t) + \int_0^t m(t-x)\, f_X(x)\, dx \tag{4.3}$$

where $m(t)$ is referred to as the **renewal function**.

Analytic evaluation of the renewal function requires the use of differential equations or Laplace transforms. Numerical analysis (or discrete-event simulation) can be used to approximate the renewal function.

For a renewal process $\{N(t), t \geq 0\}$, we can also derive a general expression for the long-run rate of renewals. This result is often referred to as the **elementary renewal theorem**.

Let $\{N(t), t \geq 0\}$ be a renewal process, and let $\{X(1), X(2), \ldots\}$ denote the corresponding sequence of inter-renewal times. If μ denotes the expected value of $X(n), n = 1, 2, \ldots$, then

$$\lim_{t \to \infty} \frac{N(t)}{t} = \frac{1}{\mu}$$

Example 4.1

Let X denote the time to failure of a unit of equipment where the PDF of X is given by

$$f_X(x) = 0.08e^{-0.08x} \quad x > 0$$

where time is measured in weeks. Upon failure, the failed unit is instantaneously replaced with an identical, independent, and new unit of equipment. What is the long-run equipment failure rate?

Let $N(t)$ denote the number of failures in $[0,t]$.

Let $X(1)$ denote the time of the first failure.

Let $X(n)$ denote the time between the $(n-1)$th and nth failure, $n = 2, 3, \ldots$.

Note that f_X denotes the probability density function of $X(n)$, $n = 1, 2, \ldots$,

$$\mu = E(X(n)) = \int_0^\infty 0.08xe^{-0.08x}\, dx$$

$$\mu = E(X(n)) = 12.5 \text{ weeks}$$

$$\text{long-run failure rate} = \lim_{t\to\infty} \frac{N(t)}{t} = \frac{1}{\mu} = 0.08 \text{ failures per week} \square$$

4.3.1 Renewal-Reward Processes

In some cases, a reward or cost is associated with each event of a renewal process. Such cases may be evaluated with the application of renewal-reward processes.

> Let $\{N(t), t \geq 0\}$ be a renewal process, and let R_n denote the reward obtained upon the nth renewal, $n = 1, 2, \ldots$. If $\{R_1, R_2, \ldots\}$ is a collection of independent and identically distributed random variables each having mean ρ, and
>
> $$R(t) = \sum_{n=1}^{N(t)} R_n$$
>
> then $\{R(t), t \geq 0\}$ is said to be a **renewal-reward process**. Furthermore,
>
> $$\lim_{t\to\infty} \frac{R(t)}{t} = \frac{\rho}{\mu}$$

Example 4.1 (continued)

A technician is contracted to take care of equipment replacements. The technician charges $7000 per replacement. What is the technician's long-run wage?

Let R_n denote the cost of the nth failure, $n = 1, 2, \ldots$

Let $R(t)$ denote the total cost incurred in $[0,t]$ where

$$R(t) = \sum_{n=1}^{N(t)} R_n$$

$$\rho = E(R_n) = \$7000$$

$$\text{long-run wage } = \lim_{t\to\infty} \frac{R(t)}{t} = \frac{\rho}{\mu} = \frac{\$7000}{12.5} = \$560 \text{ per week}\square$$

Example 4.1 (continued)

Another technician has offered to begin managing equipment replacements. This technician charges \$5000 if a replacement occurs within 8 weeks of the previous replacement, and \$7500 if the replacement occurs more than 8 weeks but less than 15 weeks after the previous replacement. If a unit of equipment survives for 15 weeks, then the technician replaces it at that point in time at a cost of \$10,000. What would be this technician's long-run average wage?

$$\mu = \int_0^{15} 0.08xe^{-0.08x}\,dx + \int_{15}^{\infty} 15\left(0.08e^{-0.08x}\right)dx = 8.735 \text{ weeks}$$

$$\rho = \int_0^{8} \$5000\left(0.08e^{-0.08x}\right)dx + \int_8^{15} \$7500\left(0.08e^{-0.08x}\right)dx + \int_{15}^{\infty} \$10000\left(0.08e^{-0.08x}\right)dx$$

$$\rho = \$7071$$

$$\text{long-run wage} \equiv \lim_{t\to\infty} \frac{R(t)}{t} = \frac{\rho}{\mu} = \frac{\$7071}{8.735} = \$809.50 \text{ per week }\square$$

4.3.2 Alternating Renewal Processes

In some applications, such as equipment maintenance planning, we consider systems that alternate between two states such that each transition corresponds to a renewal event.

Let $\{W(t), t \geq 0\}$ be a continuous-time stochastic process that alternates between two states: 0 and 1. Let $Z(n)$ denote the duration of the nth occupation of state 0, and let $Y(n)$ denote the duration of the nth occupation of state 1, $n = 1, 2, \ldots$. Suppose $\{Z(1), Z(2), \ldots\}$ is a collection of independent and identically distributed random variables, and suppose $\{Y(1), Y(2), \ldots\}$ is a collection of independent and identically distributed random variables. Furthermore, suppose the two sequences are mutually independent. Suppose the process is initially in state 0, and let

$$X(n) = Z(n) + Y(n)$$

$n = 1, 2, \ldots$. If $\{N(t), t \geq 0\}$ is a renewal process having the sequence of inter-renewal times $\{X(1), X(2), \ldots\}$, then $\{N(t), t \geq 0\}$ is said to be an alternating **renewal process.**

Furthermore, if

$$p_0 = \lim_{t \to \infty} \Pr\left[W(t) = 0\right]$$

and

$$p_1 = \lim_{t \to \infty} \Pr\left[W(t) = 1\right]$$

then,

$$p_0 = \frac{E\left[Z(n)\right]}{E\left[Z(n)\right] + E\left[Y(n)\right]}$$

and

$$p_1 = \frac{E\left[Y(n)\right]}{E\left[Z(n)\right] + E\left[Y(n)\right]}$$

Example 4.1 (continued)

Suppose that instead of being instantaneous, each equipment replacement consumes an average of one week. Assuming the original technician is used, what is the long-run availability of the equipment?

Let $W(t) = 0$ if the equipment is functioning at time t.

Let $W(t) = 1$ if the equipment is being replaced at time t.

Let $Z(n)$ denote the duration of the nth occupation of state 0.

Let $Y(n)$ denote the duration of the nth occupation of state 1.

long-run availability

$$= p_0 = \frac{E[Z(n)]}{E[Z(n)] + E[Y(n)]} = \frac{12.5}{12.5+1} = 0.9259 = 92.59\% \; \square$$

4.4 Bernoulli Processes

We now consider our first specific class of stochastic processes, the set of discrete-time counting processes known as a Bernoulli process. This class of stochastic processes is based on the simple concept of a Bernoulli trial.

Consider a random experiment having two possible outcomes that are referred to as "success" and "failure." If p denotes the probability of success, then the random experiment is referred to as a **Bernoulli trial** having success probability p. If $X = 1$ indicates the trial is a success, and $X = 0$ indicates the trial is a failure, then X is said to be a **Bernoulli random variable** with success probability p. This fact is denoted by $X \sim \text{Bernoulli}(p)$. Furthermore,

$$E(X) = p$$

$$Var(X) = p(1-p)$$

$$M_X(s) = 1 - p + pe^s$$

and

$$A_X(s) = 1 - p + ps$$

Note that any random experiment with two outcomes can be classified as a Bernoulli trial. Furthermore, the definition of success and failure is arbitrary and not tied to the colloquial connotations associated with the words *success* and *failure*.

The study of Bernoulli trials becomes more interesting when a sequence of trials is performed.

Consider a sequence of independent Bernoulli trials with each trial having probability of success p. If $N(t)$ denotes the number of successes in the first t trials, then $\{N(t), t = 1, 2, \ldots\}$ is said to be a **Bernoulli process**. This fact is denoted by $\{N(t), t = 1, 2, \ldots\} \sim \text{BP}(p)$. Furthermore, $\{N(t),$

$t = 1, 2, ...$ } is a counting process having independent increments and stationary increments, and $N(t) \sim \text{bin}(t,p)$, $t = 1, 2, \ldots$.

The fact that a Bernoulli process is a counting process is somewhat obvious, since $N(t)$ counts the number of successes that occur. The fact that a Bernoulli process has independent increments is a result of the assumed independence of the Bernoulli trials. The fact that a Bernoulli process has stationary increments is a result of the assumption that each trial has the same probability of success.

The following examples are representative of the types of situations typically modeled using Bernoulli processes.

1. Suppose a manufacturing process is such that 2% of products are defective. Let $N(t)$ denote the number of the first t items produced that are defective.
2. Suppose 16% of all customers at a restaurant order iced tea. Let $N(t)$ denote the number of the first t customers who order iced tea.

Given a fixed number of independent and identical Bernoulli trials, the binomial probability distribution describes the number of successes. We can also model the number of trials required to achieve a fixed number of successes. This model is a realization of the time-frequency duality relationship described in section 4.3.

Consider a sequence of independent Bernoulli trials with each trial having probability of success p. If $T(k)$ denotes the number of trials until the kth success where $k \in \{1, 2, \ldots\}$, then $T(k)$ is a **negative binomial random variable** with parameters k and p. This fact is denoted by $T(k) \sim$ negbin (k, p). Furthermore, the range of $T(k)$ is $\{k, k+1, \ldots\}$, the probability mass function of $T(k)$ is

$$f_{T(k)}(t) = \binom{t-1}{k-1} p^k (1-p)^{t-k}$$

$t = k, k+1, \ldots,$

$$E(T(k)) = k/p$$

$$Var(T(k)) = \frac{k(1-p)}{p^2}$$

$$M_{T(k)}(s) = \left[\frac{pe^s}{1-(1-p)e^s}\right]^k$$

and

$$A_{T(k)}(s) = \left[\frac{ps}{1 - s(1-p)} \right]^k$$

The number of independent and identical Bernoulli trials required to achieve one success ($k = 1$) is often referred to as a **geometric random variable** with parameter p.

To simplify our presentation, we use the following notation relative to computing point and cumulative binomial and negative binomial probabilities.

$$b(x, n, p) = \binom{n}{x} p^x (1-p)^{n-x}$$

$$B(x, n, p) = \sum_{j=0}^{x} b(j, n, p)$$

$$b^{-1}(k, x, p) = \binom{x-1}{k-1} p^k (1-p)^{x-k}$$

$$B^{-1}(k, x, p) = \sum_{j=k}^{x} b^{-1}(k, j, p)$$

Example 4.2

Suppose we inspect a sequence of items from a production process, and each item is classified as either conforming to technical specifications or nonconforming. The probability that any individual item is nonconforming is $p = 0.03$.

(a) How would you model the occurrence of nonconforming items?

$N(t)$ = number of nonconforming items in the first t items inspected

$\{N(t), t = 1, 2, \ldots\} \sim \text{BP}(0.03)$

(b) What is the probability that two of the first 100 items inspected will be nonconforming?

$$\Pr[N(100) = 2] = b(2, 100, 0.03) = \binom{100}{2} (0.03)^2 (0.97)^{98} = 0.2252$$

(c) What is the probability that less than two of the first 100 items inspected will be nonconforming?

$$\Pr\left[N(100)<2\right]=\Pr\left[N(100)\le 1\right]=B(1,100,0.03)=\sum_{j=0}^{1}b(j,100,0.03)=0.1946$$

(d) On average, how many of the first 100 items inspected will be nonconforming?

$$E\left[N(100)\right]=100(0.03)=3$$

(e) What is the probability that the first nonconforming item will be discovered with the 78th inspection?

$$\Pr\left[T(1)=78\right]=b^{-1}(1,78,0.03)=0.03(0.97)^{77}=0.0029$$

(f) On average, when will the first nonconforming item be discovered?

$$E\left[T(1)\right]=\frac{1}{0.03}=33.\bar{3}\text{ items into inspection}$$

(g) What is the probability that the third nonconforming item will be discovered with the 134th inspection?

$$\Pr\left[T(3)=134\right]=b^{-1}(3,134,0.03)=\binom{133}{2}(0.03)^{3}(0.97)^{131}=0.0044$$

(h) What is the probability that the third nonconforming item will not be discovered until after the 134th inspection?

$$\Pr\left[T(3)>134\right]=1-\Pr\left[T(3)\le 134\right]=1-B^{-1}(3,134,0.03)$$

$$\Pr\left[T(3)>134\right]=1-\sum_{j=3}^{134}b^{-1}(3,j,0.03)=0.2307$$

(i) On average, when will the third nonconforming item be discovered?

$$E\left[T(3)\right] = \frac{3}{0.03} = 100 \text{ items into inspection}$$

For parts (j) to (m), suppose two of the first 75 items inspected are nonconforming.

(j) What is the probability that five of the first 200 items inspected will be nonconforming?

$$\Pr\left[N(200) = 5 \middle| N(75) = 2\right] = \Pr\left[N(125) = 3\right] = b(3, 125, 0.03) = 0.2087$$

(k) On average, how many of the first 150 items inspected will be nonconforming?

$$E\left[N(150) \middle| N(75) = 2\right] = 2 + E\left[N(75)\right] = 2 + 75(0.03) = 4.25$$

(l) What is the probability that the fourth nonconforming item will be found on the 200th inspection?

$$\Pr\left[T(4) = 200 \middle| N(75) = 2\right] = \Pr\left[T(2) = 125\right] = b^{-1}(2, 125, 0.03) = 0.0026$$

(m) On average, when will the sixth nonconforming item be found?

$$E\left[T(6) \middle| N(75) = 2\right] = 75 + E\left[T(4)\right] = 75 + \frac{4}{0.03} = 208.\bar{3} \text{ items into inspection } \square$$

Homework Problems

4.1 Introduction to Stochastic Processes

(1) Imagine that you are observing a warehouse owned by a manufacturer of home electronics. In the problems that follow, do not use the same example more than once.

(a) Identify a discrete-time, discrete-valued stochastic process.

(b) Identify a discrete-time, continuous-valued stochastic process.

(c) Identify a continuous-time, discrete-valued stochastic process.

(d) Identify a continuous-time, continuous-valued stochastic process.

4.2 Introduction to Counting Processes

(1) Imagine that you are observing a university bookstore on the first day of class. Identify two counting processes. Comment on whether or not you think these counting processes have independent or stationary increments.

4.3 Introduction to Renewal Processes

(1) Let X denote the time to failure of a unit of equipment where the PDF of X is given by

$$f_X(x) = 0.0001e^{-0.0001x} \quad x > 0$$

where time is measured in hours. Upon failure, the failed unit is instantaneously replaced with an identical, independent, and new unit of equipment.

(a) What is the long-run equipment failure rate?

(b) A technician is contracted to take care of equipment replacements. The technician charges \$20,000 per replacement. What is the technician's long-run wage?

(c) The technician is considering switching to a policy under which a unit of equipment is replaced upon failure or after 1500 hours of operation. Under this new policy, if the equipment replacement results from a failure, then the technician's fee is \$12,000. If the equipment replacement occurs prior to failure, then the technician's fee is \$23,000. Under this new policy, what is the technician's long-run wage?

(d) Suppose that instead of being instantaneous, each equipment replacement consumes an average of 40 hours. Assuming the new replacement policy is used, what is the long-run availability of the equipment?

(2) Let X denote the time to failure of a unit of equipment where the CDF of X is given by

$$F_X(x) = \begin{cases} 0 & x \le 0 \\ 1 - e^{-0.01x}(1 + 0.01x) & x > 0 \end{cases}$$

where time is measured in days. Upon failure, the equipment is instantaneously restored to an "as good as new" condition.

(a) What is the long-run equipment failure rate?

If the unit of equipment operates without failure for 30 days, the unit is shut down and preventative maintenance (PM) is performed. Suppose PM is instantaneous and restores the unit to a "good as new" condition.

(b) What is the long-run rate at which maintenance is performed?

(c) Suppose a repair costs \$100, but a PM action only costs \$25. What is the long-run maintenance cost rate for the equipment?

Suppose maintenance is not instantaneous. In fact, suppose the time required to complete repair is a random variable with a mean of 2 days and the time required to complete PM is a random variable having a mean of 0.5 day.

(d) If the PM policy is not implemented, what is the long-run proportion of time that the equipment is functioning?

(e) If the PM policy is implemented, what is the long-run proportion of time that the equipment is functioning?

(f) If the PM policy is implemented, what is the long-run maintenance cost rate for the equipment?

(3) Verify Equations (4.1) and (4.2).

(4) Verify both equalities in Stement (4.3).

4.4 Bernoulli Processes

(1) Imagine that you are observing a personal computer assembly facility. Identify two Bernoulli processes.

(2) A library imposes fines for overdue books, and 12% of all customers returning books to the circulation desk have overdue books. Assume that all customers having overdue books pay their fines immediately.

(a) How would you appropriately model the collection of fines?

(b) What is the probability that three of the first twelve customers must pay a fine?

(c) What is the probability that the second fine is paid by the ninth customer?

(d) If three of the first ten customers pay fines, how many of the first 25 customers, on average, would you expect to pay fines?

(3) An electronics manufacturer subjects a certain type of integrated circuit to thermal testing. The integrated circuits are tested sequentially, and only 2% of the integrated circuits fail during the test. Those that survive the test are sent to the assembly area for inclusion in personal computers.

(a) How would you appropriately model the testing of integrated circuits?

(b) On average, how many of the first 100 integrated circuits to be tested are used in assembly?

(c) The research and development department studies those integrated circuits that do not survive the testing. On average, how many tests will be performed before the research and development department has five integrated circuits to study?

(d) Given that the 12th integrated circuit is the first one to fail, what is the probability that three of the first twenty tests result in failure?

(4) As people enter a shopping mall, they are asked to fill out a survey. Only 60% of people agree to fill out the survey. Of those people that fill out the survey, 27% respond "no" to question number 3.

(a) How would you appropriately model responses to question number 3?

(b) If six of the first 25 people to enter the mall respond "no" to question number 3, what is the probability that 17 of the first 100 people to enter the mall respond "no" to question number 3?

(c) If the seventh person to enter the mall is the second person to respond "no" to question number 3, what is the probability that the tenth person to respond "no" to question number 3 is the 81st person to enter the mall?

(d) On average, how many people will enter the mall until the first "no" response to question number 3 is obtained?

(5) Customers are supposed to rewind video tapes before returning them to a video store, and 96% of customers comply with this request. If a customer does not rewind a video tape, the store employees must rewind the tape. Assume that each customer returns a single video tape.

(a) How would you model the required rewinding of video tapes by employees?

(b) What is the probability that the employees do not have to rewind a video tape until the sixth customer?

(c) If two of the first twenty customers do not rewind their video tape, how many customers, on average, will return video tapes until employees must rewind a tape for the fifth time?

(d) If two of the first twenty customers do not rewind their video tape, and 200 customers return video tapes, how many times, on average, will the employees be required to rewind a video tape?

(6) Consider a Bernoulli process with $p = 0.3$. Determine the following.

(a) $E\left[N(3) \mid N(2) \le 1\right]$

(b) $E\left[T(1) \mid T(1) > 3\right]$

(c) $\Pr\left[T(3) = 7 \mid T(1) \le 3\right]$

Application: Acceptance Sampling

In quality control, manufacturers use various statistical methods to ensure the quality of their manufactured products. Some of these methods are designed to verify the quality of components purchased from outside suppliers. **Acceptance sampling** is such a method.

In most industrial processes, some of the items used as inputs to the process are received **in lots** from suppliers. Of these supplied items, some types may be sufficiently critical that the quality of each item in the lot is checked. In other words, lots of these items are subjected to **100% inspection**. Other types of items may not be so critical. In addition, the inspection of some types of items may be very costly or even destructive to the item. For these types of items, we may choose to perform **no inspection**. However, the more common procedure is to utilize acceptance sampling to decide whether or not the quality of the items in the received lot is adequate.

Acceptance sampling typically consists of taking a **random sample** of items from the lot, **inspecting** and classifying the quality of each item in the sample, and then, based on the information obtained via the sample, either **accepting** the lot for use in the process, or **rejecting** the lot and returning it to the supplier.

Most traditional acceptance sampling **plans** utilize an **attributes quality measure**. With an attributes measure, each inspected item is classified as either **conforming** or **nonconforming**. Thus, the quality of a received lot can be described as follows.

Let N denote number of items in the lot. Let p denote proportion of items in the lot that are nonconforming. The quantity p is referred to as the **proportion nonconforming** or the **lot quality**. Note that a larger value of p corresponds to lower quality.

Acceptance sampling plan performance is measured in two ways relative to lot quality: (1) how well does it discriminate between good and poor quality, and (2) how much inspection is required. The result is two acceptance sampling plan performance measures:

1. the OC (operating characteristic) function,
2. the ASN (average sample number) function

The **OC function** is a function of lot quality and defined as

$$P_a(p) = \Pr(\text{lot is accepted given the lot quality is } p).$$

The sample size for an acceptance sampling plan may be a fixed quantity. However, it may also be a variable that depends on what happens during the inspection process. Thus, we can also define the ASN Function as

$$ASN(p) = E(\text{number of items inspected given the lot quality is } p).$$

The first type of attributes acceptance sampling **strategy** we will consider is **single sampling**. A single sampling **plan** operates as follows:

Step 1: Inspect a random sample of n items from the lot.

$$\text{Let } X_i = \begin{cases} 0 & \text{if item } i \text{ is conforming} \\ 1 & \text{if item } i \text{ is nonconforming} \end{cases}$$

Step 2: If

$$\sum_{i=1}^{n} X_i \le c$$

accept the lot. Otherwise, reject the lot.

A single sampling plan is completely defined or identified by specifying n and c. Note that n and c are positive integers such that $n < N$ and $c < n$. The phrase "single sampling" refers to the single sample that is used to make the decision regarding **lot disposition**. The lot disposition rule can be restated as:

Accept the lot if c or fewer nonconforming items are found in the sample.

Task 1: Construct the OC function for a single sampling plan in terms of n, c, and p. In doing so, assume that $N >> n$.

Task 2: Plot the OC function for a single sampling plan having $n = 100$ and $c = 2$ over the range $p \in [0,0.1]$.

Task 3: Construct the ASN function for a single sampling plan in terms of n, c, and p.

Task 4: Plot the ASN function for a single sampling plan having $n = 100$ and $c = 2$ over the range $p \in [0,0.1]$.

A **double sampling plan** operates as follows:

Step 1: Inspect a random sample of n_1 items from the lot.

$$\text{Let } X_i = \begin{cases} 0 & \text{if item } i \text{ is conforming} \\ 1 & \text{if item } i \text{ is nonconforming} \end{cases}$$

Step 2: If

$$\sum_{i=1}^{n_1} X_i \le c_1$$

accept the lot and stop. If

$$\sum_{i=1}^{n_1} X_i > c_2$$

reject the lot and stop. Otherwise, continue.

Step 3: Inspect a random sample of an additional n_2 items from the lot.

$$\text{Let } X_i = \begin{cases} 0 & \text{if item } i \text{ is conforming} \\ 1 & \text{if item } i \text{ is nonconforming} \end{cases}$$

Step 4: If

$$\sum_{i=1}^{n_1+n_2} X_i \le c_2$$

accept the lot. Otherwise, reject the lot.

Task 5: Construct the OC function for a double sampling plan in terms of n_1, n_2, c_1, c_2, and p.

Task 6: Plot the OC function for a double sampling plan having $n_1 = 50$, $n_2 = 100$, $c_1 = 1$, and $c_2 = 4$ over the range $p \in [0,0.15]$.

Task 7: Construct the ASN function for a double sampling plan in terms of n_1, n_2, c_1, c_2, and p.

Task 8: Plot the ASN function for a double sampling plan having $n_1 = 50$, $n_2 = 100$, $c_1 = 1$, and $c_2 = 4$ over the range $p \in [0,0.3]$.

Suppose we implement the above double sampling plan ($n_1 = 50$, $n_2 = 100$, $c_1 = 1$, $c_2 = 4$). Suppose that two nonconforming items are found in the first sample. Therefore, we continue and take a second sample. Suppose the ninth item in the second sample is nonconforming, the 27th item in the second sample is nonconforming, and the 34th item in the second sample is nonconforming.

At this point, we have observed a total of five nonconforming items. Regardless of what happens during the remainder of the second sample, the ultimate decision is going to be to reject the lot. Some organizations choose to stop inspecting at this point. Such a practice is called double sampling with **curtailment**.

There are two important points regarding curtailment.

1. We do not curtail during the first sample.
2. We do not curtail when the decision to accept becomes implied.

Curtailment has no impact on the OC function (lot disposition). However, it definitely impacts the ASN function.

Task 9: Construct the ASN function for a double sampling plan with curtailment in terms of n_1, n_2, c_1, c_2, and p.

Task 10: Plot the ASN function for a double sampling plan with curtailment having $n_1 = 50$, $n_2 = 100$, $c_1 = 1$, and $c_2 = 4$ over the range $p \in [0,0.3]$.

Task 11: Construct a plot that combines the plots from Tasks 9 and 10. Comment on the savings resulting from curtailment.

5

Poisson Processes

We now dedicate an entire chapter to the most widely recognized stochastic process, the Poisson process. The Poisson process is a continuous-time counting process, and it is also a renewal process.

5.1 Introduction to Poisson Processes

We begin our discussion of the Poisson process by defining a Poisson random variable.

A discrete random variable X having range $\{0, 1, \ldots\}$ is said to be a **Poisson random variable** having mean α, that is, $X \sim \text{Poisson}(\alpha)$, if and only if the PMF of X is given by

$$f_X(x) = \frac{e^{-\alpha}\alpha^x}{x!}$$

$x = 0, 1, \ldots$. Furthermore,

$$E(X) = \alpha$$

$$Var(X) = \alpha$$

$$M_X(s) = e^{\alpha(e^s - 1)}$$

and

$$A_X(s) = e^{\alpha(s-1)}$$

You may recognize the form of the Poisson probability distribution from several examples and homework problems from earlier chapters.

Having defined a Poisson random variable, we can now define the Poisson process.

A continuous-time counting process $\{N(t), t \geq 0\}$ is said to be a **Poisson process** having rate λ, $\lambda > 0$, that is, $\{N(t), t \geq 0\} \sim \text{PP}(\lambda)$, if and only if

1. $N(0) = 0$
2. $\{N(t), t \geq 0\}$ has independent increments
3. the number of events in any interval $(s, s + t]$ where $s \geq 0$ and $t \geq 0$ is a Poisson random variable with mean λt

Part (3) of this definition implies that a Poisson process has stationary increments.

The parameter λ is called the rate of the Poisson process. Its definition corresponds precisely to its connotation—λ refers to the average rate at which events occur. If we are attempting to model a sequence of occurrences of an event in the real world, and the average event occurrence rate is not approximately constant over time (events tend to occur more frequently during certain periods of time), then the Poisson process would be an inappropriate model.

Another interesting fact about the Poisson process is that the definition of a Poisson process eliminates the possibility of two or more simultaneous events.

Example 5.1

Customers arrive at a facility according to a Poisson process with rate $\lambda = 2$ customers per minute. Suppose we begin observing the facility at some point in time.

(a) What is the probability that eight customers arrive during a 5-minute interval?

Let $N(t)$ denote the number of customers arriving in t minutes.

$$\{N(t), t \geq 0\} \sim \text{PP}(2)$$

$$\lambda t = 2(5) = 10$$

$$N(5) \sim \text{Poisson}(10)$$

$$\Pr\left(N(5) = 8\right) = \frac{e^{-10}(10)^8}{8!} = 0.1126$$

(b) On average, how many customers will arrive during a 3.2-minute interval?

$$E\left(N(3.2)\right) = 2(3.2) = 6.4$$

(c) What is the probability that more than two customers arrive during a 1-minute interval?

$$\Pr\left(N(1)>2\right)=1-\Pr\left(N(1)\le 2\right)=1-\left(\frac{e^{-2}2^0}{0!}+\frac{e^{-2}2^1}{1!}+\frac{e^{-2}2^2}{2!}\right)=0.3233$$

(d) What is the probability that four customers arrive during the interval that begins 3.3 minutes after we start observing and ends 6.7 minutes after we start observing?

$$\Pr\left(N(6.7)-N(3.3)=4\right)=\Pr\left(N(3.4)=4\right)=\frac{e^{-6.8}(6.8)^4}{4!}=0.0992$$

(e) On average, how many customers will arrive during the interval that begins 16 minutes after we start observing and ends 17.8 minutes after we start observing?

$$E\left(N(17.8)-N(16)\right)=E\left(N(1.8)\right)=2(1.8)=3.6$$

(f) What is the probability that seven customers arrive during the first 12.2 minutes we observe, given that five customers arrive during the first 8 minutes?

$$\Pr\left(N(12.2)=7 \middle| N(8)=5\right)=\Pr\left(N(4.2)=2\right)=\frac{e^{-8.4}(8.4)^2}{2!}=0.0079$$

(g) If three customers arrive during the first 1.2 minutes of our observation period, then, on average, how many customers will arrive during the first 3.7 minutes?

$$E\left(N(3.7) \middle| N(1.2)=3\right)=3+E\left(N(2.5)\right)=3+2(2.5)=8$$

(h) If one customer arrives during the first 6 seconds of our observations, then what is the probability that two customers arrive during the interval that begins 12 seconds after we start observing and ends 30 seconds after we start observing?

$$\Pr\left(N(0.5)-N(0.2)=2 \middle| N(0.1)=1\right)=\Pr\left(N(0.5)-N(0.2)=2\right)=\Pr\left(N(0.3)=2\right)$$

$$\Pr\left(N(0.5)-N(0.2)=2 \middle| N(0.1)=1\right)=\frac{e^{-0.6}(0.6)^2}{2!}=0.0988$$

(i) If five customers arrive during the first 30 seconds of our observations, then, on average, how many customers will arrive during the interval that begins 1 minute after we start observing and ends 3 minutes after we start observing?

$$E\left(N(3)-N(1) \middle| N(0.5)=5\right)=E\left(N(3)-N(1)\right)=E\left(N(2)\right)=2(2)=4$$

(j) If three customers arrive during the interval that begins 1 minute after we start observing and ends 2.2 minutes after we start observing, then, on average, how many customers will arrive during the first 3.7 minutes?

$$E\left(N(3.7) \middle| N(2.2)-N(1)=3\right)=E\left(N(3.7) \middle| N(1.2)=3\right)=$$

$$3+E\left(N(2.5)\right)=3+2(2.5)=8$$

□

5.2 Interarrival Times

Although the number of events that occur during some time interval may be of great interest when studying a counting process, it may also be of interest to study the timing of events.

Let $\{N(t), t \geq 0\}$ be a continuous-time counting process. If $S(n)$ denotes the time at which the nth event occurs, $n = 1, 2, \ldots$, then $S(n)$ is referred to as the arrival time of the nth event, and $\{S(n), n = 1, 2, \ldots\}$ is the sequence of **arrival times** for the counting process. If $T(1) = S(1)$ and $T(n) = S(n) - S(n-1)$, $n = 2, 3, \ldots$, then $T(n)$ is referred to as the **interarrival time** of the nth event, and $\{T(n), n = 1, 2, \ldots\}$ is the sequence of interarrival times for the counting process.

The probabilistic behavior of the interarrival times of a Poisson process is one of the most well-known results from the theory of stochastic processes.

The continuous-time counting process $\{N(t), t \geq 0\}$ is a Poisson process having rate λ, if and only if $\{T(n), n = 1, 2, \ldots\}$ is a sequence of independent and identically distributed exponential random variables having rate λ.

This result is another realization of the time-frequency duality relationship. The result implies the need for a formal definition of the exponential random variable and exploration of some of its interesting properties. This result also implies that Poisson processes are renewal processes.

A continuous random variable X having range $(0, \infty)$ is said to be an **exponential random variable** having rate α, $\alpha > 0$, that is, $X \sim \text{expon}(\alpha)$, if and only if the PDF of X is given by

$$f_X(x) = \alpha e^{-\alpha x},\ x \in (0, \infty)$$

If $X \sim \text{expon}(\alpha)$, then the CDF of X is given by

$$F_X(x) = \begin{cases} 0 & x \le 0 \\ 1 - e^{-\alpha x} & x > 0 \end{cases}$$

$$E(X) = 1/\alpha$$

$$Var(X) = 1/\alpha^2$$

and

$$M_X(s) = \frac{\alpha}{\alpha - s}$$

You may also recognize the form of the exponential probability distribution from several examples and homework problems from earlier chapters.

Example 5.1 (continued)

(k) What is the probability that the third customer arrives more than 0.2 but less than 1.3 minutes after the second customer?

$$\Pr(0.2 < T(3) < 1.3) = \Pr(T(3) < 1.3) - \Pr(T(3) < 0.2)$$

$$T(3) \sim \text{expon}(2)$$

$$\Pr(0.2 < T(3) < 1.3) = \left(1 - e^{-2(1.3)}\right) - \left(1 - e^{-2(0.2)}\right) = 0.5960$$

(l) What is the expected value of the time between the first and second customer arrivals?

$$E\left(T\left(2\right)\right)=\frac{1}{2}=0.5\text{ min}$$

(m) If the time between the first and second customer arrivals is 0.6 minute, then what is the probability that it is more than 1.7 minutes between the third and fourth customer arrivals?

Since $T(2)$ and $T(4)$ are independent,

$$\Pr\left(T\left(4\right)>1.7\middle|T\left(2\right)=0.6\right)=\Pr\left(T\left(4\right)>1.7\right)=e^{-2(1.7)}=0.0334\ \square$$

One of the more interesting properties of an exponential random variable is the **memoryless property**. Furthermore, the memoryless property uniquely defines the exponential random variable among all continuous random variables.

A continuous random variable X is an exponential random variable if and only if

$$\Pr\left(X>s+t\mid X>s\right)=\Pr\left(X>t\right) \tag{5.1}$$

for all $s\in(0,\infty)$, and for all $t\in(0,\infty)$. If equation (5.1) holds, then

$$E\left(X\mid X>s\right)=s+E\left(X\right)=s+\frac{1}{\alpha} \tag{5.2}$$

for all $s\in(0,\infty)$.

There are two common situations in which the memoryless property calls into question the application of the exponential random variable and thus the application of the Poisson process. First, if an exponential random variable is used to model the lifetime of a device, then at every point in time until it fails, the device is as good as new. If the exponential random variable is used to model a customer arrival time, then the arrival of the customer does not become more imminent over time.

Example 5.1 (continued)

(n) Suppose it has been 1.3 minutes since the tenth customer arrived. What is the probability that it is less than 0.8 minute until the eleventh customer arrives?

$$\Pr\left(T\left(11\right)<2.1\middle|T\left(11\right)>1.3\right)=\Pr\left(T\left(11\right)<0.8\right)=1-e^{-2(0.8)}=0.7981$$

(o) Suppose it has been 0.7 minute since the sixteenth customer arrived. What is the expected value of the seventeenth customer's interarrival time?

$$E\left(T(17)\middle|T(17)>0.7\right)=0.7+E\left(T(17)\right)=0.7+\frac{1}{2}=1.2\ \text{min}\square$$

Two additional results regarding the exponential random variable involve the consideration of multiple, independent exponential random variables. These random variables are sometimes referred to as **competing exponentials.**

Suppose X_1 is an exponential random variable having rate α_1 and X_2 is an exponential random variable having rate α_2. If X_1 and X_2 are independent, then

$$\Pr\left(X_1 < X_2\right)=\frac{\alpha_1}{\alpha_1+\alpha_2} \tag{5.3}$$

If $\{X_1, X_2, \dots, X_n\}$ is a collection of independent random variables such that $X_j \sim \text{expon}(\alpha_j)$, $j = 1, 2, \dots, n$, and $Y = \min(X_1, X_2, \dots, X_n)$, then $Y \sim \text{expon}(\alpha)$ where

$$\alpha=\sum_{j=1}^{n}\alpha_j \tag{5.4}$$

Example 5.2

Two people, Kevin and Bob, leave for work at the same time. Kevin's commute requires an amount of time that is exponentially distributed with a mean of 10 minutes. Bob's commute is independent of Kevin's and requires an exponentially distributed amount of time with a mean of 7 minutes. What is the probability that Bob arrives at work before Kevin?

Let X_1 denote Kevin's commuting time.

Let X_2 denote Bob's commuting time.

$$X_1 \sim \text{expon}(1/10)$$

$$X_2 \sim \text{expon}(1/7)$$

$$\Pr\left(X_2 < X_1\right)=\frac{\alpha_2}{\alpha_2+\alpha_1}=\frac{1/7}{1/7+1/10}=0.5882\,\square$$

Example 5.3

A store takes orders from three independent types of customers. The time until an order of type i is received is an exponential random variable with mean $1/\alpha_i$ hours where $\alpha_1 = 0.2$, $\alpha_2 = 0.5$, and $\alpha_3 = 0.05$. What is the probability that it is more than 1 hour until an order is received?

Let X_i denote time until an order of type i is received.

Let Y denote the time until an order is received.

$$X_i \sim \text{expon}(\alpha_i)$$

$$Y = \min(X_1, X_2, X_3) \sim \text{expon}(0.75)$$

$$\Pr(Y > 1) = e^{-0.75(1)} = 0.4724 \;\square$$

5.3 Arrival Times

For a Poisson process $\{N(t),\, t \geq 0\}$, the CDF of $S(n)$ can be obtained through algebraic manipulation of the CDF of $N(t)$. The result is another recognizable form.

If $\{N(t),\, t \geq 0\}$ is a Poisson process having rate λ, then $S(n)$ is an n-Erlang random variable having rate λ, $n = 1, 2, \ldots$.

This result necessitates the need for formal definition of the Erlang random variable and exploration of some of its interesting properties.

A continuous random variable X having range $(0, \infty)$ is said to be an **r-Erlang random variable**, $r \in \{1, 2, \ldots\}$, having rate α, $\alpha > 0$, that is, $X \sim r\text{-Erlang}(\alpha)$, if and only if the PDF of X is given by

$$f_X(x) = \frac{\alpha^r x^{r-1} e^{-\alpha x}}{(r-1)!}, \quad x \in (0, \infty)$$

If $X \sim r\text{-Erlang}(\alpha)$, then the CDF of X is given by

$$F(x) = \begin{cases} 0 & x \leq 0 \\ 1 - e^{-\alpha x} \displaystyle\sum_{j=0}^{r-1} \frac{(\alpha x)^j}{j!} & x > 0 \end{cases}$$

$$E(X) = r/\alpha$$

$$Var(X) = r/\alpha^2$$

and

$$M_X(s) = \left(\frac{\alpha}{\alpha - s}\right)^r$$

The r-Erlang random variable is a special case of what is commonly known as a gamma random variable, and the 1-Erlang random variable is equivalent to an exponential random variable.

Example 5.1 (continued)

(p) What is the probability that the fourth customer arrives before we have completed 2.5 minutes of observation?

$$S(4) \sim \text{4-Erlang}(2)$$

$$\Pr\left(S(4) < 2.5\right) = 1 - e^{-5}\sum_{j=0}^{3}\frac{5^j}{j!} = 0.7350$$

Any probability question about $S(n)$ can be converted to an equivalent expression about $N(t)$.

$$\Pr\left(S(4) < 2.5\right) = \Pr\left(N(2.5) \geq 4\right) = 1 - \Pr\left(N(2.5) \leq 3\right)$$

(q) On average, how long will it be until the third customer arrives?

$$E\left(S(3)\right) = \frac{3}{2} = 1.5 \text{ min}$$

(r) If the fourth customer arrives 2.3 minutes after we begin observing, then what is the probability that the sixth customer arrives after we have completed 4 minutes of observation?

$$\Pr\left(S(6) > 4 \mid S(4) = 2.3\right) = \Pr\left(S(2) > 1.7\right) = e^{-3.4}\sum_{j=0}^{1}\frac{3.4^j}{j!} = 0.1468$$

(s) If the second customer arrives 1.4 minutes after we begin observing, then what is the expected value of the fourth customer's arrival time?

$$E\left(S(4)\middle|S(2)=1.4\right)=1.4+E\left(S(2)\right)=1.4+\frac{2}{2}=2.4 \text{ min}$$

(t) If the second customer arrives more than 0.8 minute after the first customer, then, on average, when will the third customer arrive?

$$E\left(S(3)\middle|T(2)>0.8\right)=0.8+E\left(S(3)\right)=2.3 \text{ min}$$

(u) If five customers arrive during the first 0.4 minute, then what is the expected value of the seventh customer's arrival time?

$$E\left(S(7)\middle|N(0.4)=5\right)=0.4+E\left(S(2)\right)=1.4 \text{ min}$$

(v) If the third customer arrives 1.2 minutes after we begin observing, then what is the probability that seven customers arrive during the first 3 minutes?

$$\Pr\left(N(3)=7\middle|S(3)=1.2\right)=\Pr\left(N(1.8)=4\right)=\frac{e^{-3.6}(3.6)^4}{4!}=0.1912 \ \square$$

If $\{N(t), t \geq 0\}$ is a counting process, then by definition

$$S(n)=\sum_{j=1}^{n} T(j)$$

This relationship leads to another general result regarding exponential and Erlang random variables.

If $\{X_1, X_2, \ldots, X_r\}$ is a collection of independent and identically distributed expon(α) random variables, and

$$Y=\sum_{j=1}^{r} X_j$$

then

$$Y \sim r\text{-Erlang}(\alpha) \tag{5.5}$$

5.4 Decomposition and Superposition of Poisson Processes

Suppose each event in a Poisson process can be classified into two categories, and suppose the classification of events can be modeled as a sequence of independent and identical Bernoulli trials. This process is often referred to as **decomposition of a Poisson process**. The following result for decomposed Poisson processes can be extended to the case in which the classification includes more than two categories.

> Let $\{N(t), t \geq 0\}$ be a Poisson process having rate λ. Suppose at the occurrence of each event a Bernoulli trial with probability of success p is performed. If $\{N_1(t), t \geq 0\}$ denotes the counting process for the number of successes and $\{N_2(t), t \geq 0\}$ denotes the counting process for the number of failures, then $\{N_1(t), t \geq 0\}$ is a Poisson process having rate λp, and $\{N_2(t), t \geq 0\}$ is a Poisson process having rate $\lambda(1-p)$. Furthermore $\{N_1(t), t \geq 0\}$ and $\{N_2(t), t \geq 0\}$ are independent (5.6)

Just as Poisson processes can be decomposed, they can also be combined. This process is often referred to as **superposition of Poisson processes**.

> Let $\{N_j(t), t \geq 0\}$ be a Poisson process having rate λ_j, $j = 1, 2, \ldots, m$. Furthermore, suppose these m Poisson processes are independent. If
>
> $$N(t) = \sum_{j=1}^{m} N_j(t)$$
>
> then $\{N(t), t \geq 0\}$ is a Poisson process having rate λ, where
>
> $$\lambda = \sum_{j=1}^{m} \lambda_j \tag{5.7}$$

There is a natural relationship between Poisson process decomposition and superposition. This relationship is best explored through the use of examples.

Example 5.4

Customers arrive at a bank according to a Poisson process having rate $\lambda = 35$ per hour, and 56% of customers are women.

(a) Describe the arrival process of female customers.

Let $N_1(t)$ denote the number of female customers to arrive in the first t hours.

$$\lambda p = 35(0.56) = 19.6$$

$$N_1(t) \sim PP(19.6)$$

(b) Given that 42 customers arrive in an hour, what is the probability that 25 of them are women?

Let $N(t)$ denote the number of customers to arrive in the first t hours.

$$\Pr\left(N_1(1) = 25 \middle| N(1) = 42\right) = b(25, 42, 0.56) = 0.1120 \ \square$$

Example 5.5

Traffic converges from four independent directions at an intersection. Cars arrive from the east according to a Poisson process having a rate of three cars per minute. Cars arrive from the south according to a Poisson process having a rate of two cars per minute. Cars arrive from the west according to a Poisson process having a rate of three cars per minute. Cars arrive from the north according to a Poisson process having a rate of one car per minute.

(a) Describe the arrival process of cars to the intersection.

Let $N_E(t)$ denote the number of cars that arrive from the east in the first t minutes.

Let $N_S(t)$ denote the number of cars that arrive from the south in the first t minutes.

Let $N_W(t)$ denote the number of cars that arrive from the west in the first t minutes.

Let $N_N(t)$ denote the number of cars that arrive from the north in the first t minutes.

$$\{N_E(t), t \geq 0\} \sim PP(3)$$
$$\{N_S(t), t \geq 0\} \sim PP(2)$$
$$\{N_W(t), t \geq 0\} \sim PP(3)$$
$$\{N_N(t), t \geq 0\} \sim PP(1)$$

Let $N(t)$ denote the number of cars that arrive in the first t minutes.

$$N(t) = N_E(t) + N_S(t) + N_W(t) + N_N(t)$$

$$\{N(t), t20\} \sim PP(9)$$

(b) Given that two cars arrive from the east during the first minute, what is the probability that eleven cars arrive during the first minute?

$$\Pr\left(N(1)=11 \middle| N_E(1)=2\right)=\Pr\left(N_S(1)+N_W(1)+N_N(1)=9\right)=\frac{e^{-6}6^9}{9!}=0.0688$$

(c) If 21 cars arrive during a 2-minute period, what is the probability that six of them were from the east?

$$\Pr\left(N_E(2)=6 \middle| N(2)=21\right)=\frac{\Pr\left(N_E(2)=6, N(2)=21\right)}{\Pr\left(N(2)=21\right)}$$

$$\Pr\left(N_E(2)=6 \middle| N(2)=21\right)=\frac{\Pr\left(N_E(2)=6, N_S(2)+N_W(2)+N_N(2)=15\right)}{\Pr\left(N(2)=21\right)}$$

$$\Pr\left(N_E(2)=6 \middle| N(2)=21\right)=\frac{\Pr\left(N_E(2)=6\right)\Pr\left(N_S(2)+N_W(2)+N_N(2)=15\right)}{\Pr\left(N(2)=21\right)}$$

$$=\frac{\left(\frac{e^{-6}6^6}{6!}\right)\left(\frac{e^{-12}12^{15}}{15!}\right)}{\frac{e^{-18}18^{21}}{21!}}=\frac{21!}{6!15!}\left(\frac{6}{18}\right)^6\left(\frac{12}{18}\right)^{15}$$

$$\Pr\left(N_E(2)=6 \middle| N(2)=21\right)=b(6,21,3/9)=0.1700$$

(d) During a given period, thirty cars arrive, five of which are from the south. What is the expected value of the number of cars arriving from the east or west during that period?

$$E\left(N_E(t)+N_W(t) \middle| N(t)=30, N_S(t)=5\right)$$

$$=E\left(N_E(t)+N_W(t) \middle| N_E(t)+N_W(t)+N_N(t)=25\right)$$

$$N_E(t)+N_W(t) \sim PP(6)$$

$$N_E(t)+N_W(t)+N_S(t) \sim PP(7)$$

$$E\left(N_E(t)+N_W(t) \middle| N(t)=30, N_S(t)=5\right)=25(6/7)=21.4 \square$$

5.5 Competing Poisson Processes

The concept of competing exponentials can be extended to competing Poisson process arrival times.

Let $\{N_1(t), t \geq 0\}$ be a Poisson process having rate λ_1, and let $\{N_2(t), t \geq 0\}$ be a Poisson process having rate λ_2. If $\{N_1(t), t \geq 0\}$ and $\{N_2(t), t \geq 0\}$ are independent, then

$$\Pr\left(S_1(n) < S_2(m)\right) = \sum_{k=n}^{n+m-1} \binom{n+m-1}{k} \left(\frac{\lambda_1}{\lambda_1 + \lambda_2}\right)^k \left(\frac{\lambda_2}{\lambda_1 + \lambda_2}\right)^{n+m-1-k} \tag{5.7}$$

for all $n = 1, 2, \ldots$, and for all $m = 1, 2, \ldots$.

Equation (5.7) provides a formula for computing the probability that the nth arrival from one Poisson process occurs before the mth arrival from a second, independent Poisson process.

Example 5.5 (continued)

Suppose we begin observing the intersection at some point in time.

(e) What is the probability that a car arrives from the east before a car arrives from the south?

$$\Pr\left(S_E(1) < S_S(1)\right) = \frac{3}{3+2} = 0.6$$

(f) What is the probability that two cars arrive from the east before a car arrives from the south?

$$\Pr\left(S_E(2) < S_S(1)\right) = \left(\frac{3}{3+2}\right)^2 = 0.36$$

(g) What is the probability that four cars arrive from the east before the third car arrives from the south?

$$\Pr\left(S_E(4) < S_S(3)\right) = \sum_{k=4}^{6} \binom{6}{k} \left(\frac{3}{5}\right)^k \left(\frac{2}{5}\right)^{6-k}$$

$$\Pr\left(S_E(4) < S_S(3)\right) = 0.5443 \;\square$$

5.6 Nonhomogeneous Poisson Processes

What if the average rate at which events occur is not constant over time? In such a case, we may use a time-dependent function, $\lambda(t)$, to describe the instantaneous rate of event occurrence at time t.

A continuous-time counting process $\{N(t), t \geq 0\}$ is said to be a **nonhomogeneous Poisson process** having **intensity function** $\lambda(t)$, where $\lambda(t) \geq 0$ for $t \in (0, \infty)$, if and only if

1. $N(0)=0$
2. $\{N(t), t \geq 0\}$ has independent increments
3. The number of events in any interval $(s, s + t]$ where $s \geq 0$ and $t \geq 0$ is a Poisson random variable with mean $[\Lambda(t+s) - \Lambda(s)]$ where $\Lambda(t)$ is referred to as the **cumulative intensity function** and given by

$$\Lambda(t) = \int_0^t \lambda(u)\, du$$

Nonhomogeneous Poisson processes do not have stationary increments, and they are not renewal processes.

Example 5.6

Suppose a machine is switched on at time $t = 0$ (where time is measured in hours). Suppose the machine's reliability and maintainability characteristics are such that machine failures occur according to a nonhomogeneous Poisson process having intensity function

$$\lambda(t) = 0.2t\,,\ t \geq 0.$$

(a) Determine the cumulative intensity function.

$$\Lambda(t) = \int_0^t 0.2u\, du = 0.1t^2 \quad t \geq 0$$

(b) Determine the probability distribution of the number of failures in the first three hours of machine use.

$$\Lambda(3) = 0.1(3)^2 = 0.9$$

$$N(3) \sim \text{Poisson}(0.9)$$

(c) Determine the probability distribution of the number of failures in the second three hours of machine use.

$$\Lambda(6) - \Lambda(3) = 0.1(6)^2 - 0.1(3)^2 = 2.7$$

$$N(6) - N(3) \sim \text{Poisson}(2.7) \square$$

Homework Problems

5.1 Introduction to Poisson Processes

(1) Customers orders are submitted to an Internet site according to a Poisson process having rate $\lambda = 3.1$ orders per hour. Suppose we begin observing the Internet site at some point in time.

(a) What is the probability that eight orders occur during a 3-hour interval?

(b) On average, how many orders will occur during a 5-hour interval?

(c) What is the probability that more than three orders occur during a 1-hour interval?

(d) What is the probability that six orders occur during the interval that begins 1 hour after we start observing and ends 4 hours after we start observing?

(e) On average, how many orders will occur during the interval that begins 2 hours after we start observing and ends 4 hours after we start observing?

(f) What is the probability that seven orders occur during the first 3.5 hours we observe, given that three orders occur during the first 1.5 hours?

(g) If nine orders occur during the first 3 hours of our observation period, then, on average, how many orders will occur during the first 13.25 hours?

(h) If two orders occur during the first 0.5 hour of our observations, then what is the probability that four orders occur during the interval that begins 1 hour after we start observing and ends 2.5 hours after we start observing?

(i) If 11 orders occur during the first 3.5 hours of our observations, then, on average, how many orders will occur during the interval that begins 5 hours after we start observing and ends 12 hours after we start observing?

(j) If six orders occur during the interval that begins 2 hours after we start observing and ends 3.5 hours after we start observing, then, on average, how many orders will occur during the first 7 hours?

(2) Cars arrive at an intersection according to a Poisson process having a rate of three cars per minute. Suppose we are interested in studying car arrivals on a particular day.

(a) What is the probability that three cars arrive at the intersection between 10:00 a.m. and 10:02 a.m.?

(b) What is the probability that more than three cars arrive at the intersection between 10:00 a.m. and 10:02 a.m.?

(c) On average, how many cars will arrive at the intersection before 6:00 p.m.?

(d) On average, how many cars will arrive at the intersection between 6:00 p.m. and 7:15 p.m.?

(e) What is the expected value of the number of cars to arrive at the intersection, given that 2127 arrive before 8:00 a.m.?

(3) Between 10:00 a.m. and 12:00 p.m., female customers arrive at a bank according to a Poisson process having a rate of 7.3 per hour.

(a) Given that twelve female customers arrive at the bank between 10:00 a.m. and 11:30 a.m., what is the probability that sixteen female customers arrive between 10:00 a.m. and 12:00 p.m.?

(b) What is the expected value of the number of female customers to arrive between 10:00 a.m. and 12:00 p.m.?

(c) Given that five female customers arrive between 10:00 a.m. and 11:45 a.m., what is the expected value of the number of female customers to arrive between 10:00 a.m. and 12:00 p.m.?

(d) Given that five female customers arrive between 10:00 a.m. and 11:45 a.m., what is the probability that least one but no more than three female customers arrive between 11:50 a.m. and 12:00 p.m.?

(4) Let $\{N(t), t \geq 0\}$ be a Poisson process having rate $\lambda = 1$. Determine the following.

(a) $\Pr[N(3.7) = 3 \mid N(2.2) \geq 2]$

(b) $\Pr[N(3.7) = 1 \mid N(2.2) < 2]$
(c) $E[N(5) \mid N(10) = 7]$

5.2 Interarrival Times

(1) Verify equation (5.1).

(2) Verify equation (5.2).

(3) Let X be an exponential random variable having rate $\alpha = 1$. Determine the following.
(a) $\Pr(X > 1.7 \mid X > 1.2)$
(b) $E(X \mid X > 0.6)$
(c) $\Pr(X \leq 2.2 \mid X > 1.3)$
(d) $E(X \mid X < 0.3)$
(e) $\Pr(X \leq 1.3 \mid X < 2)$

(4) Consider a Poisson process $\{N(t), t \geq 0\}$. Verify that $T(1) \sim \text{expon}(\lambda)$.

(5) Customer orders are submitted to an Internet site according to a Poisson process having rate $\lambda = 3.1$ orders per hour. Suppose we begin observing the Internet site at some point in time.
(a) What is the probability that the fourth order occurs more than 15 minutes but less than 30 minutes after the third order?
(b) What is the expected value of the time between the fourth and fifth orders?
(c) If the time between the seventh and eighth orders is 45 seconds, then what is the probability that it is more than 10 minutes between the tenth and eleventh orders?
(d) Suppose it has been 12 minutes since the seventh order occurred. What is the probability that it is less than 15 minutes until the eighth order occurs?
(e) Suppose it has been 20 minutes since the third order occurred. What is the expected value of the interarrival time of the fourth order?

(6) During a certain time period, passenger arrivals at a train station can be modeled as a Poisson process having a rate of $\lambda = 4.6$ passengers per minute.
(a) What is the expected value of the elapsed time between the third and fourth passenger arrivals?

(b) What is the probability that the elapsed time between the third and fourth passenger arrivals is less than 15 seconds?

(c) It has been 30 seconds since the second passenger arrived, and the third passenger has yet to arrive. What is the expected value of the interarrival time of the third passenger?

(7) Suppose the occurrence of major earthquakes in a certain part of southern California can be modeled as a Poisson process having a rate of 0.35 earthquake per year.

(a) What is the probability that it is at least 2 years before the next major earthquake?

(b) It has been 0.71 year since the last major earthquake. What is the probability that it will be at least 6 but no more than 9 months before the next major earthquake?

(c) It has been 1.4 years since the last major earthquake. On average, how long will it be until the next major earthquake?

(8) Verify equation (5.3).

(9) Verify equation (5.4).

(10) A facility contains three machines. The time until a machine fails is an exponential random variable. For the first machine, the rate is 0.1. For the second, the rate is 0.12, and for the third, the rate is 0.2. Assuming the time units are hours, determine the following.

(a) What is the probability that machine 1 fails before machine 2?

(b) What is the probability that no machine fails for at least 6.5 hours?

(c) What is the probability that machine 3 is the first to fail?

(11) An emergency dispatching service receives three types of calls (fire, police, medical). The time until each type of call is an exponential random variable. For fire calls, the rate is 0.5. For police calls, the rate is 4, and for medical calls, the rate is 7. Assuming the time units are hours, determine the following.

(a) What is the probability that a medical call occurs before a fire call?

(b) What is the probability that a call is received within 6 minutes?

(c) What is the probability that the first call is a fire call?

5.3 Arrival Times

(1) For a Poisson process $\{N(t), t \geq 0\}$, verify the form of the CDF of $S(n)$ through algebraic manipulation of the CDF of $N(t)$.

(2) Customers orders are submitted to an Internet site according to a Poisson process having rate $\lambda = 3.1$ orders per hour. Suppose we begin observing the Internet site at some point in time.

(a) What is the probability that the third order occurs before we have completed 1 hour of observation?

(b) On average, how long will it be until the sixth order occurs?

(c) If the sixth order occurs 1.25 hours after we begin observing, then what is the probability that the tenth order occurs after we have completed 90 minutes of observation?

(d) If the fifth order occurs 2 hours after we begin observing, then what is the expected value of the arrival time of the ninth order?

(e) If the fourth order occurs more than 15 minutes after the third order, then, on average, when will the seventh order occur?

(f) If eight orders occur during the first 2.25 hours, then what is the expected value of the 12th order's arrival time?

(g) If the fifth order occurs 1.2 hours after we begin observing, then what is the probability that nine orders occur during the first 3 hours?

(3) During a certain time period, passenger arrivals at a train station can be modeled as a Poisson process having a rate of $\lambda = 4.6$ passengers per minute.

(a) If we start observing at 7:00 p.m., then what is the expected value of the time of the eighth passenger arrival?

(b) If we start observing at 7:00 p.m. and at 7:30 p.m., 124 passengers have arrived, then what is the expected value of the arrival time of the 140th passenger?

(4) Suppose the occurrence of major earthquakes in a certain part of southern California can be modeled as a Poisson process having a rate of 0.35 earthquake per year.

(a) Suppose we start observing earthquakes now, and the first major earthquake does not occur for at least 4 years. On average, how long will it be until the fifth major earthquake?

(b) Suppose we start observing earthquakes now, and the third major earthquake occurs 3.5 years from now. What is the probability that five major earthquakes occur in the next 10 years?

(c) Suppose we start observing earthquakes now, and the fifth major earthquake occurs 8.2 years from now. What is the probability that the seventh major earthquake occurs within the next 10 years?

(5) Verify statement (5.5).

5.4 Decomposition and Superposition of Poisson Processes

(1) Verify statement (5.6).

(2) Verify statement (5.7).

(3) Cars arrive at a fast-food restaurant according to a Poisson process having a rate of $\lambda = 34$ cars per hour. Suppose 21% of cars proceed to the drive-in service area.

(a) What is the probability that nine cars proceed to the drive-in area in a 30-minute period?

(b) During a one-hour period, 14 cars went to the drive-in area. What is the expected value of the number of cars that arrived at the restaurant during that one-hour period?

(c) What is the probability that it is at least 12 minutes between the arrivals of the fourth and sixth cars at the drive-in area?

(d) Of the cars that do not proceed to the drive-in service area, 3% leave the restaurant without stopping. On average, how many such cars will arrive during a 90-minute period?

(4) E-mail messages arrive at a server according to a Poisson process having a rate of $\lambda = 1.5$ messages per second. Suppose 40% of messages are blocked by the server's spam filter.

(a) What is the probability that seven messages are blocked during a 10-second period?

(b) During a 30-second period, 23 messages were blocked. What is the expected value of the number of messages that arrived at the server during that period?

(c) What is the probability that it is at least 1 second between the third and sixth blocked messages?

(d) Of the messages that are not blocked, 10% are rejected due to an unknown address. On average, how many such messages arrive during a 1-minute period?

(5) During a certain time period, students arrive at a university library according to a Poisson process having a rate of 4.5 per minute. During that same time period, faculty arrive according to a Poisson process having a rate of 0.8 per minute, and all other people (we refer to them as the general public) arrive according to a Poisson process having a rate of 2.1 per minute.

(a) What is the probability that three people arrive at the library in a one-minute interval?

(b) Given that three faculty arrive during a two-minute interval, what is the expected value of the number of people to arrive during a 5-minute interval that completely contains that 2-minute interval?

(c) What is the expected value of the number of students and faculty to enter the library during a 15-minute interval, given that 27 members of the general public arrived during that same interval?

(d) Given that 12 people entered the library during a certain time interval, what is the probability that 6 of them were students?

(6) The customer service center for an airline's frequent flier program receives telephone calls from customers in three tiers of their frequent flier program. Tier 1 customers call according to a Poisson process having a rate of 1.2 per minute. Tier 2 customers call according to a Poisson process having a rate of 2.3 per minute. Tier 3 customers call according to a Poisson process having a rate of 4.1 per minute.

(a) What is the probability that 12 calls are received during a 1.5-minute interval?

(b) Given that four tier 1 customers call during a 2-minute interval, what is the expected value of the number of calls during a 6-minute interval that completely contains that 2-minute interval?

(c) What is the expected value of the number of tier 3 calls received during a 20-minute interval, given that 47 tier 2 calls were received during that same interval?

(d) Given that 20 calls are received during a certain time interval, what is the probability that seven of them were from tier 2 customers?

5.5 Competing Poisson Processes

(1) Verify equation (5.8).

(2) Undergraduate students arrive at an academic building according to a Poisson process having a rate of 10.5 students per hour. Graduate students arrive at the building according to a Poisson process having a rate of 4.5 per hour. Suppose we begin observing the building at some point in time.

(a) What is the probability that the first student to arrive is a graduate student?

(b) What is the probability that the first two students to arrive are undergraduates?

(c) What is the probability that the tenth undergraduate student arrives before the fifth graduate student?

(3) The customer service center for an airline's frequent flier program receives telephone calls from customers in three tiers of their frequent flier program. Tier 1 customers call according to a Poisson process having a rate of 1.2 per minute. Tier 2 customers call according to a Poisson process having a rate of 2.3 per minute. Tier 3 customers call according to a Poisson process having a rate of 4.1 per minute. Suppose we begin monitoring calls at some point in time.

(a) What is the probability that the first call is from a tier 1 customer?

(b) What is the probability that the first three calls are from tier 3 customers?

(c) What is the probability that the eighth tier 2 call is received before the fifth tier 1 call?

5.6 Nonhomogeneous Poisson Processes

(1) Over a 10-hour period, customers arrive at a restaurant according to a nonhomogeneous Poisson process having intensity function

$$\lambda(t) = 50 - 2(t-5)^2 \quad 0 \le t \le 10$$

where time is measured in hours.

(a) Find the probability distribution of the number of customer arrivals during the first two hours.

(b) Find the probability distribution of the number of customer arrivals during the second two hours.

(2) A store opens at 8:00 a.m. From 8:00 a.m. until 10:00 a.m., customers arrive at an average rate of four per hour. Between 10:00 a.m. and noon, they arrive at an average rate of eight per hour. From noon until 2:00 p.m., the average arrival rate increases linearly from eight per hour to ten per hour. From 2:00 p.m. until 5:00 p.m., the average arrival rate decreases linearly from ten per hour to four per hour. Note: This problem is adapted from Ross (2000).
   (a) Determine the probability distribution on the number of customers who enter the store on a given day.
   (b) Determine the probability distribution on the number of customers who enter the store between 8:00 a.m. and noon on a given day.
   (c) Determine the probability distribution on the number of customers who enter the store between noon and 5:00 p.m. on a given day.

Application: Repairable Equipment

A company uses a piece of equipment in their manufacturing process that is subject to failure. The time to failure for the unit is a Weibull random variable having shape parameter $\beta > 0$ and scale parameter $\eta > 0$. In other words, if T denotes the time to failure of a new unit and F_T is the CDF of T, then

$$F_T(t) = 1 - e^{-(t/\eta)^\beta} \quad t > 0$$

This piece of equipment is the bottleneck of the continuous (24 hours per day, 7 days per week) manufacturing process. Thus when it fails, minimal repair actions are taken. In other words, the maintenance technicians do just enough to restore the unit to a functioning state.

Under the minimal repair model, equipment failures occur according to a nonhomogeneous Poisson process having intensity function $z(t)$ where $z(t)$ corresponds to the hazard function of T. The hazard function of T is given by

$$z(t) = \frac{f_T(t)}{1 - F_T(t)}$$

where f_T denotes the PDF of T.

Task 1: Construct a general expression for the hazard function of a Weibull time to failure distribution.

Task 2: Let $N(t)$ denote the number of failures in the first t hours of equipment function. Describe the probability distribution of $N(t)$.

Suppose $\beta = 2.2$ and $\eta = 300$ hours.

Task 3: What is the probability of more than 30 equipment failures during the first 1000 hours of equipment function? During the second 1000 hours of equipment function?

Task 4: Over time, does the equipment's reliability improve, stay the same, or degrade? Why?

The company has a preventive maintenance policy that states:

> After τ hours of equipment *function*, the unit is shut down and overhauled.

The intention of an overhaul is to restore the unit to an "as good as new" condition. Suppose each repair causes the company to incur a cost of c_r and each overhaul results in an incurred cost of c_o.

Task 5: Assuming both types of maintenance actions (repair and overhaul) are instantaneous, construct a general expression for C, the unit's long-run maintenance cost per hour.

Task 6: Construct a general expression for τ^*, the optimal overhaul policy.

Suppose $c_r = \$100$ and $c_o = \$2500$.

Task 7: Compute the optimal overhaul policy for the unit. Use a plot to verify that it is the optimal value.

Task 8: Modify this optimal policy into an easy-to-implement, but near-optimal, practical policy. How much money would the company lose over the course of one year by implementing the practical policy?

Suppose maintenance actions are not instantaneous. Suppose the time required to complete a repair is t_r and the time required to complete an overhaul is t_o.

Task 9: Construct a general expression for A, the long-run availability of the unit.

Task 10: Construct a general expression for τ^*, the optimal overhaul policy (relative to availability).

Suppose $t_r = 1$ hour and $t_o = 15$ hours.

Task 11: Compute the optimal overhaul policy for the unit. Use a plot to verify that it is the optimal value. What is the optimal unit availability?

Task 12: If the policy identified in Task 11 is implemented, how many equipment failures would you expect to observe during a 5-year period?

6

Discrete-Time Markov Chains

There are many physical, economic, and managerial systems for which historical information is not relevant in predicting on-going behavior. Among those systems, those that can be modeled using a stochastic process are subject to a special approach to modeling and analysis.

A stochastic process for which the future behavior depends only on the current state of the process is called a Markov chain. In this chapter, we explore the discrete-time, discrete-valued subset of such processes—discrete-time Markov chains.

6.1 Introduction

Suppose a system can be reasonably modeled by a series of transitions among a countable number of states. In addition, suppose these transitions occur at discrete points in time. Furthermore, suppose the sequence of states occupied in the future can be described probabilistically using only the identity of the current state. Under such a scenario, we can use a discrete-time Markov chain to model the behavior of the system.

> If $\{X(t), t = 0, 1, \ldots\}$ is a discrete-valued stochastic process having a countable state space K, then $\{X(t), t = 0, 1, \ldots\}$ is said to be a **discrete-time Markov chain** if and only if for all $\{i_0, i_1, \ldots, i_{t-1}, i, j\} \subseteq K$ and for all $t = 0, 1, \ldots,$
>
> $$\Pr\left(X(t+1) = j \mid X(t) = i,\ X(t-1) = i_{t-1},\ \ldots, X(1) = i_1,\ X(0) = i_0\right)$$
> $$= \Pr\left(X(t+1) = j \mid X(t) = i\right) = P_{ij}$$
>
> This property is referred as the **Markov property**. The value P_{ij} is referred to as the **transition probability** from state i to state j.

The Markov property can be stated colloquially as:

Given the present, the future is independent of the past.

By stating that a stochastic process possesses the Markov property, we are stating that the probability that the process occupies some state in the future only depends on our knowledge of the most recent state.

We usually summarize the behavior of a discrete-time Markov chain by putting the transition probabilities in matrix form.

Let $\{X(t), t = 0, 1, \ldots\}$ be a discrete-time Markov chain having state space K. Let **P** denote the matrix of transition probabilities where the row corresponds to the current state and the column corresponds to the next state. The matrix **P** is referred to as the **transition probability matrix**. Note that

$$P_{ij} \geq 0$$

for all $i \in K, j \in K$. Furthermore,

$$\sum_{j \in K} P_{ij} = 1$$

for all $i \in K$.

For example, if the discrete-time Markov chain has state space $\{1, 2, 3\}$, then the transition probability matrix is

$$\mathbf{P} = \begin{bmatrix} P_{11} & P_{12} & P_{13} \\ P_{21} & P_{22} & P_{23} \\ P_{31} & P_{32} & P_{33} \end{bmatrix}$$

We know that the probabilities in each row of the transition probability matrix must sum to 1 (the process has to transition to *some* state). If the columns also sum to 1, then we refer to the transition matrix and the Markov chain as doubly stochastic.

Let $\{X(t), t = 0, 1, \ldots\}$ be a discrete-time Markov chain having state space K. The discrete-time Markov chain is referred to as **doubly stochastic** if

$$\sum_{i \in K} P_{ij} = 1$$

for all $j \in K$.

We note some special properties of doubly stochastic discrete-time Markov chains later in this chapter.

In order to model a system as a discrete-time Markov chain, we must be able to define the state space and timing mechanism (for transitions) in a meaningful way such that the Markov property holds. Having done that, the next step is to construct or describe the transition probability matrix.

Example 6.1

Each item produced by a manufacturing process is either conforming to specifications or nonconforming. If an item is conforming, then the probability that the next item is nonconforming is 0.01. If an item is nonconforming, then the probability that the next item is nonconforming is 0.1. Model this production process using a discrete-time Markov chain.

Let $X(t) = 0$ if the tth item produced is conforming, $t = 1, 2, \ldots$.

Let $X(t) = 1$ if the tth item produced is nonconforming, $t = 1, 2, \ldots$.

$$\mathbf{P} = \begin{bmatrix} 0.99 & 0.01 \\ 0.9 & 0.1 \end{bmatrix} \square$$

Example 6.2

An organization has employees at three levels (level 1, level 2, level 3). At the end of each year, 13% of level 1 employees are promoted to level 2 and 18% of level 1 employees leave the organization; 21% of level 2 employees are promoted to level 3, and 17% of level 2 employees leave the organization; and 9% of level 3 employees leave the organization. Demotions never occur. Model the status of an employee using a discrete-time Markov chain.

$$\text{Let } X(t) = \begin{cases} 1 & \text{if the employee is level 1 at the end of year } t \\ 2 & \text{if the employee is level 2 at the end of year } t \\ 3 & \text{if the employee is level 3 at the end of year } t \\ 4 & \text{if the employee is not in the organization at the end of year } t \end{cases} \qquad t = 0, 1, \ldots$$

$$\mathbf{P} = \begin{bmatrix} 0.69 & 0.13 & 0 & 0.18 \\ 0 & 0.62 & 0.21 & 0.17 \\ 0 & 0 & 0.91 & 0.09 \\ 0 & 0 & 0 & 1 \end{bmatrix} \square$$

Example 6.3

A person is standing in a certain location. With probability p, she takes one step to the left. Otherwise, she takes one step to the right. Model this process, the **random walk**, using a discrete-time Markov chain.

Suppose step locations are numbered using integers such that a step to the left corresponds to a unit decrease in the integer, and a step to the right corresponds to a unit increase in the integer. Let K denote the set of integers.

Let $X(t)$ denote the person's location after t steps, $t = 0, 1, \ldots$.

For all $i \in K$,

$$P_{i,i-1} = p$$

and

$$P_{i,i+1} = 1 - p \,\square$$

Example 6.4

A gambler has a certain amount of money. He places a series of \$1 bets. He stops when he has \$C or when he runs out of money. The probability of winning each bet is p. Model the gambler's progress using a discrete-time Markov chain.

Let $X(t)$ denote the gambler's total after t bets, $t = 0, 1, \ldots$.

$$P_{0,0} = P_{C,C} = 1$$

$$P_{i,i+1} = p \quad i = 1, 2, \ldots, C - 1$$

$$P_{i,i-1} = 1 - p \quad i = 1, 2, \ldots, C - 1 \,\square$$

Example 6.5

A customer orders 0, 1, or 2 items every week. The number he orders one week directly affects the number he orders the next week. The number he orders is a binomial random variable with $n = 2$. If he ordered 0 last week, then $p = 0.9$. If he ordered 1, then $p = 0.5$. If he ordered 2, then $p = 0.1$. Model the customer's ordering behavior using a discrete-time Markov chain.

Let $X(t)$ denote the number of items ordered in week t, $t = 1, 2, \ldots$.

$$\mathbf{P} = \begin{bmatrix} b(0,2,0.9) & b(1,2,0.9) & b(2,2,0.9) \\ b(0,2,0.5) & b(1,2,0.5) & b(2,2,0.5) \\ b(0,2,0.1) & b(1,2,0.1) & b(2,2,0.1) \end{bmatrix} \square$$

Example 6.6

Suppose we consider Example 6.5 under slightly different conditions. Suppose the number he orders this week depends on the total of his orders the previous two weeks.

Total Order Last Two Weeks	p
0, 1	0.9
2, 3	0.5
4	0.1

Let $X(t) = (i, j)$ if the customer orders i items in week t and j items in week $t - 1, t = 2, 3, \ldots$.

$$P_{(i,j)(i',i)} = b(i', 2, 0.9) \qquad \text{if } i + j \le 1$$

$$P_{(i,j)(i',i)} = b(i', 2, 0.5) \qquad \text{if } 2 \le i + j \le 3$$

$$P_{(i,j)(i',i)} = b(i', 2, 0.1) \qquad \text{if } i + j = 4 \;\square$$

6.2 Manipulating the Transition Probability Matrix

Once the transition probability matrix has been constructed, we can examine the dynamic behavior of the system. We can determine which states are likely to be occupied the most often, the probability that the various states are occupied at specific points in time, and the number of steps before the system will return to any state. We obtain these results through the analysis of the transition probability matrix.

Let $\{X(t), t = 0, 1, \ldots\}$ be a discrete-time Markov chain having state space K and transition probability matrix **P**. The probability

$$P_{ij}^{(n)} = \Pr\left(X(t+n) = j \mid X(t) = i\right)$$

is referred to as the ***n*-step transition probability** from state i to state j for all $i \in K, j \in K, t = 0, 1, \ldots, n = 1, 2, \ldots$. Note that

$$P_{ij}^{(1)} = P_{ij}$$

We represent all of the n-step transition probabilities at the same time using the **n-step transition probability matrix, $\mathbf{P}^{(n)}$**.

Example 6.7

Consider a discrete-time Markov chain having state space {1, 2, 3} with transition probability matrix **P** given by

$$\mathbf{P} = \begin{bmatrix} 0.2 & 0.5 & 0.3 \\ 0.6 & 0.2 & 0.2 \\ 0.3 & 0.2 & 0.5 \end{bmatrix}$$

If the process is in state 2 now, then what is the probability that it is in state 3 two periods from now?

$$P_{23}^{(2)} = P_{21}P_{13} + P_{22}P_{23} + P_{23}P_{33} = 0.6(0.3) + 0.2(0.2) + 0.2(0.5) = 0.32 \; \square$$

The general expression for this probability calculation is known as the set of Chapman-Kolmogorov equations. This set of simultaneous equations applies to all numbers of steps.

If $\{X(t), t = 0, 1, \dots\}$ is a discrete-time Markov chain having state space K and transition probability matrix **P**, then

$$P_{ij}^{(m+n)} = \sum_{k \in K} P_{ik}^{(m)} P_{kj}^{(n)}$$

for all $i \in K, j \in K, m = 1, 2, \dots, n = 1, 2, \dots$.

Conceptually, the Chapman-Kolmogorov equations imply that to transition from one state to another in $(m + n)$ number of steps, a process may first make a transition to any intermediate state in the first m steps and then continue on to the destination state in the remaining n steps. All feasible paths of this sort must be included in the calculation of the $(m + n)$-step transition probabilities.

Examination of the summation of products in the Chapman-Kolmogorov equations reveals that the calculation is really a matrix multiplication.

Let $\{X(t), t = 0, 1, \dots\}$ be a discrete-time Markov chain having state space K and transition probability matrix **P**. In general,

$$\mathbf{P}^{(m+n)} = \mathbf{P}^{(m)}\,\mathbf{P}^{(n)}$$

for all $m = 1, 2, \dots, n = 1, 2, \dots$. Note that a useful special case of this form is:

$$\mathbf{P}^{(n)} = \mathbf{P}^{(n-1)}\,\mathbf{P}$$

$n = 2, 3, \ldots$.

Example 6.7 (continued)

Determine the 2-step transition probability matrix, the 3-step transition probability matrix and $P_{31}^{(4)}$.

$$\mathbf{P}^{(2)} = \mathbf{PP} = \begin{bmatrix} 0.2 & 0.5 & 0.3 \\ 0.6 & 0.2 & 0.2 \\ 0.3 & 0.2 & 0.5 \end{bmatrix}\begin{bmatrix} 0.2 & 0.5 & 0.3 \\ 0.6 & 0.2 & 0.2 \\ 0.3 & 0.2 & 0.5 \end{bmatrix} = \begin{bmatrix} 0.43 & 0.26 & 0.31 \\ 0.30 & 0.38 & 0.32 \\ 0.33 & 0.29 & 0.38 \end{bmatrix}$$

$$\mathbf{P}^{(3)} = \mathbf{P}^{(2)}\mathbf{P} = \begin{bmatrix} 0.43 & 0.26 & 0.31 \\ 0.30 & 0.38 & 0.32 \\ 0.33 & 0.29 & 0.38 \end{bmatrix}\begin{bmatrix} 0.2 & 0.5 & 0.3 \\ 0.6 & 0.2 & 0.2 \\ 0.3 & 0.2 & 0.5 \end{bmatrix} = \begin{bmatrix} 0.335 & 0.329 & 0.336 \\ 0.384 & 0.290 & 0.326 \\ 0.354 & 0.299 & 0.347 \end{bmatrix}$$

The Chapman-Kolmogorov equations imply three ways to compute $P_{31}^{(4)}$.

$$P_{31}^{(4)} = P_{31}^{(3)}P_{11} + P_{32}^{(3)}P_{21} + P_{33}^{(3)}P_{31} = 0.354(0.2) + 0.299(0.6) + 0.347(0.3) = 0.3543$$

$$P_{31}^{(4)} = P_{31}^{(2)}P_{11}^{(2)} + P_{32}^{(2)}P_{21}^{(2)} + P_{33}^{(2)}P_{31}^{(2)} = 0.33(0.43) + 0.29(0.30) + 0.38(0.33) = 0.3543$$

$$P_{31}^{(4)} = P_{31}P_{11}^{(3)} + P_{32}P_{21}^{(3)} + P_{33}P_{31}^{(3)} = 0.3(0.335) + 0.2(0.384) + 0.5(0.354) = 0.3543$$

We could have computed $P_{31}^{(4)}$ by computing $\mathbf{P}^{(4)}$ and then noting the first element in the third row of that matrix. □

For some of the systems that we might model using discrete-time Markov chains, the starting ($t = 0$) state might also be reasonably treated as a random variable. Even when this is not the case, the dynamic behavior of the model implies that there is a probability distribution on the identity of the state occupied at any point in time. In order to determine those step-dependent distributions, we must know the distribution on state occupancy at the start of the analysis.

Let $\{X(t), t = 0, 1, \ldots\}$ be a discrete-time Markov chain having state space K and transition probability matrix $\mathbf{P}$. The probability

$$\alpha_i = \Pr\left(X(0) = i\right)$$

is referred to as the **initial occupancy probability** of state i for all $i \in K$. The vector

$$\alpha = \begin{bmatrix} \alpha_1 & \alpha_2 & \cdots \end{bmatrix}$$

is referred to as the **initial occupancy vector.** The probability

$$v_j^{(n)} = \Pr\left(X(n) = j\right)$$

is referred to as the ***n*-step occupancy probability** of state j for all $j \in K$, $n = 1, 2, \ldots$. The vector

$$v^{(n)} = \begin{bmatrix} v_1^{(n)} & v_2^{(n)} & \cdots \end{bmatrix}$$

is referred to as the ***n*-step state occupancy vector,** $n = 1, 2, \ldots$. Note that

$$v_j^{(n)} = \sum_{i \in K} \alpha_i P_{ij}^{(n)}$$

for all $j \in K$, $n = 1, 2, \ldots$. Note also that

$$v^{(n)} = \alpha\, \mathbf{P}^{(n)}$$

for all $n = 1, 2, \ldots$.

Example 6.7 (continued)

Suppose $\alpha = [0.4 \;\; 0.3 \;\; 0.3]$. Determine the 3-step occupancy vector.

$$v_1^{(3)} = \alpha_1 P_{11}^{(3)} + \alpha_2 P_{21}^{(3)} + \alpha_3 P_{31}^{(3)} = 0.4(0.335) + 0.3(0.384) + 0.3(0.354) = 0.3554$$

$$v_2^{(3)} = \alpha_1 P_{12}^{(3)} + \alpha_2 P_{22}^{(3)} + \alpha_3 P_{32}^{(3)} = 0.4(0.329) + 0.3(0.290) + 0.3(0.299) = 0.3083$$

$$v_3^{(3)} = 1 - (0.3554 + 0.3083) = 0.3363$$

$$v^{(3)} = \begin{bmatrix} 0.3554 & 0.3083 & 0.3363 \end{bmatrix} \square$$

We can also consider the probability that a discrete-time Markov chain reaches some specified state for the first time in some specified number of transitions. For example, we may wish to know the probability that an inventory of items is depleted for the first time in 14 days. Such probabilities are referred to as first passage probabilities.

Let $\{X(t), t = 0, 1, \ldots\}$ be a discrete-time Markov chain having state space K and transition probability matrix $\mathbf{P}$. The probability

$$f_{ij}^{(n)} = \Pr\left(X(m+n) = j, X(t) \neq j, t = m+1, m+2, \ldots, m+n-1 \mid X(m) = i\right)$$

is referred to as the ***n*-step first passage probability** from state i to state j for all $i \in K, j \in K, m = 0, 1, \ldots, n = 1, 2, \ldots$. The first passage probability denotes the probability that n transitions are required to move from state i to state j for the first time. Note that

$$f_{ij}^{(1)} = P_{ij}$$

Example 6.7 (continued)

Determine the 2-step and 3-step first passage probabilities from state 1 to state 2.

$$f_{12}^{(2)} = P_{11}P_{12} + P_{13}P_{32} = 0.2(0.5) + 0.3(0.2) = 0.16$$

$$f_{12}^{(3)} = P_{11}P_{11}P_{12} + P_{11}P_{13}P_{32} + P_{13}P_{31}P_{12} + P_{13}P_{33}P_{32} = 0.107 \ \square$$

Computing first passage probabilities is a straightforward process for small values of n. As n increases, it becomes increasingly difficult to conceptualize all the paths from state i to state j in n steps such that j is not reached until the nth step. Therefore, we most often compute first passage probabilities using a recursive relationship.

If $\{X(t), t = 0, 1, \ldots\}$ is a discrete-time Markov chain having state space K and transition probability matrix $\mathbf{P}$, then

$$f_{ij}^{(n)} = P_{ij}^{(n)} - \sum_{k=1}^{n-1} f_{ij}^{(k)} P_{jj}^{(n-k)}$$

for all $i \in K, j \in K, n = 1, 2, \ldots$.

Example 6.7 (continued)

Determine the 4-step first passage probability from state 1 to state 2.

$$f_{12}^{(4)} = P_{12}^{(4)} - \sum_{k=1}^{3} f_{12}^{(k)} P_{22}^{(4-k)} = P_{12}^{(4)} - \left[f_{12}^{(1)} P_{22}^{(3)} + f_{12}^{(2)} P_{22}^{(2)} + f_{12}^{(3)} P_{22}^{(1)}\right]$$

$$f_{12}^{(4)} = P_{12}^{(4)} - \left[0.5(0.290) + 0.16(0.38) + 0.107(0.2)\right]$$

$$P_{12}^{(4)} = P_{11}^{(3)}P_{12} + P_{12}^{(3)}P_{22} + P_{13}^{(3)}P_{32} = 0.3005$$

$$f_{12}^{(4)} = 0.0733 \square$$

We can also determine the expected value of the number of transitions to move from state i to state j for the first time.

Let $\{X(t), t = 0, 1, \ldots\}$ be a discrete-time Markov chain having state space K and transition probability matrix **P**. The value m_{ij} denotes the expected value of the number of transitions required to move from state i to state j for the first time for all $i \in K, j \in K$. The value m_{ij} is also referred to as the **mean first passage time** from state i to state j.

By definition,

$$m_{ij} = \sum_{n=1}^{\infty} n f_{ij}^{(n)}$$

However, from a computational standpoint, this equation is not very useful. A more useful approach is to determine all the mean first passage times for a discrete-time Markov chain simultaneously. This computational process is accomplished using a set of linear equations resulting from conditioning on the first transition of the Markov chain.

If $\{X(t), t = 0, 1, \ldots\}$ is a discrete-time Markov chain having state space K and transition probability matrix **P**, then

$$m_{ij} = 1 + \sum_{\substack{k \neq j \\ k \in K}} P_{ik} m_{kj} \tag{6.1}$$

for all $i \in K, j \in K$.

Example 6.8

Consider a discrete-time Markov chain having state space {1, 2} with transition probability matrix **P** given by

$$\mathbf{P} = \begin{bmatrix} 0.6 & 0.4 \\ 0.1 & 0.9 \end{bmatrix}$$

Determine all the mean first passage times for this process.

$$m_{11} = 1 + P_{12}m_{21}$$

$$m_{12} = 1 + P_{11}m_{12} \Rightarrow (1 - P_{11})m_{12} = 1 \Rightarrow m_{12} = 1/(1 - 0.6) = 2.5$$

$$m_{21} = 1 + P_{22}m_{21} \Rightarrow (1 - P_{22})m_{21} = 1 \Rightarrow m_{21} = 1/(1 - 0.9) = 10$$

$$m_{22} = 1 + P_{21}m_{12}$$

$$m_{11} = 1 + 0.4(10) = 5$$

$$m_{22} = 1 + 0.1(2.5) = 1.25$$ □

6.3 Classification of States

In order to consider the various types of systems modeled using discrete-time Markov chains, we must define different classifications for the states and thus different classifications for the overall stochastic process.

> Let $\{X(t), t = 0, 1, \ldots\}$ be a discrete-time Markov chain having state space K and transition probability matrix **P**. Consider a pair of states $i \in K$, and $j \in K$. We say that state j is **accessible** from state i if and only if there exists some $n \in \{1, 2, \ldots\}$ such that
>
> $$P_{ij}^{(n)} > 0$$
>
> States i and j are said to **communicate** if state i is accessible from state j and state j is accessible from state i. Note that communication is transitive, that is, if state i communicates with state j, and state j communicates with state k, then state i communicates with state k.

We can use the concept of communication to divide the state space into disjoint classes.

> Let $\{X(t), t = 0, 1, \ldots\}$ be a discrete-time Markov chain having state space K and transition probability matrix **P**. A **class** of states is a subset of K such that: (1) all states in the subset communicate with one another, and (2) no state not in the subset communicates with any state in the subset. The discrete-time Markov chain is said to be **irreducible** if it has only one class. Otherwise, the chain is said to be reducible.

Transition diagrams facilitate the identification of classes by representing the discrete-time Markov chain as a network of nodes and directed arcs. In a transition diagram, each state in a discrete-time Markov chain is represented using a node and each directed arc represents a possible one-step transition.

Example 6.1 (continued)

Identify each class of the defined Markov chain.
classes: {1,2} □

Example 6.2 (continued)

Identify each class of the defined Markov chain.
classes: {1}, {2}, {3}, {4} □

Example 6.4 (continued)

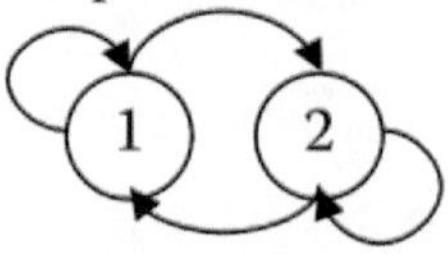

Suppose $C = 4$. Identify each class of the defined Markov chain.
classes: {0}, {1,2,3}, {4} □

Each class of states in a discrete-time Markov chain falls into one of two

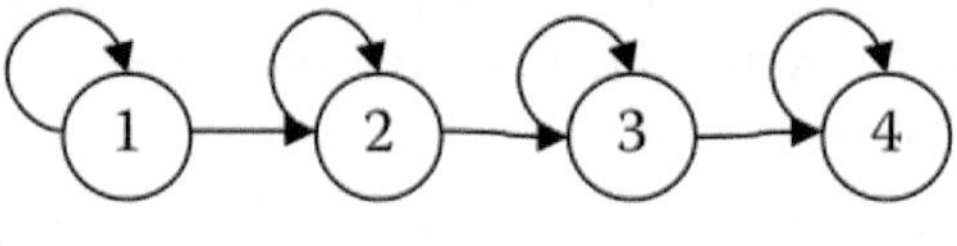

categories.

A **recurrent** class of states in a discrete-time Markov chain is a class such that the process cannot transition out of the class. If a recurrent class

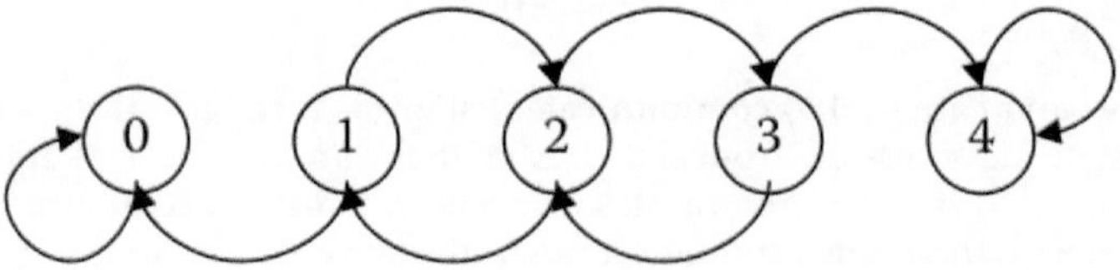

contains a single state, then the state is said to be **absorbing**.

All discrete-time Markov chains with finite state spaces have at least one recurrent class. Irreducible discrete-time Markov chains have a single recurrent class. Discrete-time Markov chains can have more than one recurrent class.

A **transient** class in a discrete-time Markov chain is a class such that if the process transitions out of the class, then there is a positive probability that the process will not reenter the class.

Example 6.1 (continued)

Identify each class of the defined Markov chain as recurrent or transient.

{1, 2}	recurrent□

Example 6.2 (continued)

Identify each class of the defined Markov chain as recurrent or transient.

{1}	transient
{2}	transient
{3}	transient
{4}	recurrent (absorbing)□

Example 6.4 (continued)

Suppose $C = 4$. Identify each class of the defined Markov chain as recurrent or transient.

{0}	recurrent (absorbing)
{1,2,3}	transient
{4}	recurrent (absorbing)□

6.4 Limiting Behavior

If a discrete-time Markov chain is irreducible and has a finite state space, then as the number of transitions increases, the initial state of the process becomes irrelevant. We summarize this behavior by saying that the process approaches **steady-state** behavior.

Let $\{X(t), t = 0, 1, \ldots\}$ be a discrete-time Markov chain having state space K and transition probability matrix **P**. Let π_j denote the **limiting** (or **stationary**) **probability** of state j where

$$\pi_j = \lim_{n \to \infty} P_{ij}^{(n)}$$

for all $i \in K, j \in K$. If

$$\pi = \begin{bmatrix} \pi_0 & \pi_1 & \cdots \end{bmatrix}$$

then π is referred to as the **limiting** (or **stationary**) **probability vector**. An equivalent alternate definition of the elements of the vector π is

$$\pi_j = \lim_{n \to \infty} v_j^{(n)}$$

As implied by the definition, we use the transition probability matrix to compute the limiting probabilities.

Let $\{X(t), t = 0, 1, \ldots\}$ be a discrete-time Markov chain having state space K and transition probability matrix $\mathbf{P}$. The vector π is the unique non-negative solution to the set of linear equations

$$\pi_j = \sum_{i \in K} \pi_i P_{ij}$$

for all $j \in K$ and

$$\sum_{j \in K} \pi_j = 1 \tag{6.2}$$

The matrix representation of the above set of equations is

$$\pi = \pi \mathbf{P}$$

The set of equations prescribed for computing the limiting probabilities includes one extra equation (as compared to the number of states of the Markov chain). The need for the extra equation, equation (6.2), results from the fact that the rows of the transition probability matrix are required to sum to one. Thus there are only $n - 1$ independent equations until we add equation (6.2). Therefore, we can arbitrarily eliminate one of the equations constructed from the transition probability matrix.

A doubly stochastic discrete-time Markov chain constitutes a special case. Let $\{X(t), t = 0, 1, \ldots\}$ be a doubly stochastic discrete-time Markov chain having a finite state space K and transition probability matrix $\mathbf{P}$. If the state space is comprised of d elements, then

$$\pi_j = \frac{1}{d}$$

for all $j \in K$.

Example 6.7 (continued)

Determine the limiting probabilities for the Markov chain.

$$\pi_1 = 0.2\pi_1 + 0.6\pi_2 + 0.3\pi_3$$

$$\pi_2 = 0.5\pi_1 + 0.2\pi_2 + 0.2\pi_3$$

$$\pi_1 + \pi_2 + \pi_3 = 1$$

$$0.8\pi_1 - 0.6\pi_2 - 0.3\pi_3 = 0$$

$$-0.5\pi_1 + 0.8\pi_2 - 0.2\pi_3 = 0$$

$$\pi_1 + \pi_2 + \pi_3 = 1$$

$$\begin{bmatrix} 0.8 & -0.6 & -0.3 \\ -0.5 & 0.8 & -0.2 \\ 1 & 1 & 1 \end{bmatrix} \begin{bmatrix} \pi_1 \\ \pi_2 \\ \pi_3 \end{bmatrix} = \begin{bmatrix} 0 \\ 0 \\ 1 \end{bmatrix}$$

$$\begin{bmatrix} \pi_1 \\ \pi_2 \\ \pi_3 \end{bmatrix} = \begin{bmatrix} 0.8 & -0.6 & -0.3 \\ -0.5 & 0.8 & -0.2 \\ 1 & 1 & 1 \end{bmatrix}^{-1} \begin{bmatrix} 0 \\ 0 \\ 1 \end{bmatrix} = \begin{bmatrix} 0.356 \\ 0.307 \\ 0.337 \end{bmatrix} \square$$

The most common interpretation of the limiting probabilities is that they represent the long-run probability distribution of the state of the Markov chain. We can apply this probability distribution as we do any other discrete probability distribution.

Example 6.1 (continued)

Interpret π_1.

The value of π_1 is the long-run probability that a produced item is nonconforming. □

Example 6.5 (continued)

Construct an expression in terms of the limiting probabilities for the long-run average order size.

$$\pi_1 + 2\pi_2 \square$$

Example 6.5 (continued)

Suppose the customer pays \$50 per item ordered plus a fixed fee of \$10 for placing an order. Construct an expression in terms of the limiting probabilities for the long-run average order cost.

$$\$60\pi_1 + \$110\pi_2 \square$$

6.5 Absorbing States

Many discrete-time Markov chains are comprised of a set of transient states and a set of more than one absorbing states. In such cases, we may be interested in determining the probability distribution associated with the absorbing state reached by the process.

Let $\{X(t), t = 0, 1, \dots\}$ be a discrete-time Markov chain having transition probability matrix **P**. Suppose the state space can be divided into a set of transient states K and a set of absorbing states J. If a_{ij} denotes the probability that the process is absorbed into state j given that the initial state is i, for all $i \in K \cup J, j \in J$, then a_{ij} is referred to as the **conditional absorption probability** of state j given an initial state i. Note that

$$\sum_{j \in J} a_{ij} = 1$$

for all $i \in K \cup J$. Furthermore, we can compute all the conditional absorption probabilities for the discrete-time Markov chain using the set of linear equations

$$a_{ij} = P_{ij} + \sum_{k \in K} P_{ik} a_{kj}$$

for all $i \in K \cup J, j \in J$.

If δ_j denotes the probability that the chain is absorbed into state j, for all $j \in J$, then δ_j is referred to as the **absorption probability** of state j, and

$$\delta_j = \alpha_j + \sum_{k \in K} \alpha_k a_{kj}$$

for all $j \in J$.

Example 6.4 (continued)

Suppose $C = 4$ and $p = 0.4$. Suppose the gambler starts with \$2.

(a) What is the probability that he stops with no money?

$$\mathbf{P} = \begin{bmatrix} 1 & 0 & 0 & 0 & 0 \\ 0.6 & 0 & 0.4 & 0 & 0 \\ 0 & 0.6 & 0 & 0.4 & 0 \\ 0 & 0 & 0.6 & 0 & 0.4 \\ 0 & 0 & 0 & 0 & 1 \end{bmatrix}$$

$$a_{20} = 0.6a_{10} + 0.4a_{30}$$

$$a_{10} = 0.6 + 0.4a_{20}$$

$$a_{30} = 0.6a_{20}$$

$$a_{20} = 0.6\left(0.6 + 0.4a_{20}\right) + 0.4\left(0.6a_{20}\right) = 0.36 + 0.48a_{20}$$

$$a_{20} = 0.692$$

(b) What is the expected value of the amount of money possessed by the gambler when he stops playing?

$$\$0\left(a_{20}\right) + \$4\left(a_{24}\right)$$

$$a_{24} = 1 - a_{20} = 0.308$$

$$\$0\left(a_{20}\right) + \$4\left(a_{24}\right) = \$1.23 \ \square$$

Example 6.4 (continued)

Suppose $C = 4$ and $p = 0.4$, and suppose the gambler has a 10% chance of beginning the game with \$1, a 30% chance of beginning the game with \$2, and a 60% chance of beginning the game with \$3. What is the probability that the gambler has \$4 when he stops playing?

$$a_{10} = 0.6 + 0.4(0.692) = 0.877 \qquad a_{14} = 0.123$$

$$a_{30} = 0.6(0.692) = 0.415 \qquad a_{34} = 0.585$$

$$\delta_4 = \alpha_1 a_{14} + \alpha_2 a_{24} + \alpha_3 a_{34} = 0.1(0.123) + 0.3(0.308) + 0.6(0.585) = 0.456 \; \square$$

We may also be interested in the expected number of transitions required to reach one of the absorbing states. Since there is no guarantee which absorbing state will be reached, we must combine the absorbing states to compute this quantity. Once the absorbing states have been combined, we can use mean first passage times to compute the expected value of interest.

Example 6.4 (continued)

Suppose C = 4 and p = 0.4, and suppose the gambler has a 10% chance of beginning the game with \$1, a 30% chance of beginning the game with \$2, and a 60% chance of beginning the game with \$3. What is the expected value of the number of bets placed by the gambler?

The absorbing states 0 and 4 are combined into a single state which we refer to as state A.

The expected value of the number of bets placed is given by

$$\alpha_1 m_{1A} + \alpha_2 m_{2A} + \alpha_3 m_{3A}$$

$$m_{1A} = 1 + 0.4 m_{2A}$$

$$m_{2A} = 1 + 0.6 m_{1A} + 0.4 m_{3A}$$

$$m_{3A} = 1 + 0.6 m_{2A}$$

$$m_{2A} = 1 + 0.6(1 + 0.4 m_{2A}) + 0.4(1 + 0.6 m_{2A}) = 2 + 0.48 m_{2A}$$

$$m_{2A} = 3.85$$

$$m_{1A} = 1 + 0.4(3.85) = 2.54$$

$$m_{3A} = 1 + 0.6(3.85) = 3.31$$

The expected value of the number of bets placed is

$$0.1(2.54) + 0.3(3.85) + 0.6(3.31) = 3.40 \square$$

If we combine the absorbing states as one, then we can take our analysis one step further. We can compute the expected value of the number of times the process visits each transient state before the process enters the absorbing state.

Let $\{X(t), t = 0, 1, \dots\}$ be a discrete-time Markov chain having transition probability matrix **P**. Suppose the state space can be divided into a set of transient states K and a single absorbing state A. Let s_{ij} denote the expected value of the number of times the process transitions into state j given that the initial state is i, for all $i \in K, j \in K$. We can compute these expected values using the set of linear equations

$$s_{ij} = P_{ij} + \sum_{k \in K} P_{ik} s_{kj}$$

for all $i \in K, j \in K$

If μ_j denotes the expected value of the number of times the process visits state j, for all $j \in K$, then

$$\mu_j = \alpha_j + \sum_{k \in K} \alpha_k s_{kj}$$

for all $j \in K$.

Example 6.2 (continued)

Suppose 50% of employees enter the organization in level 1, 30% enter the organization in level 2, and 20% enter in level 3. On average, how many years does an employee finish in each level?

$$s_{11} = 0.69 + 0.69 s_{11} + 0.13 s_{21} + 0.18 s_{41} = 0.69 + 0.69 s_{11}$$

$$s_{11} = 2.23$$

$$s_{12} = 0.13 + 0.69 s_{12} + 0.13 s_{22} + 0.18 s_{42} = 0.13 + 0.69 s_{12} + 0.13 s_{22}$$

$$s_{22} = 0.62 + 0.62 s_{22} + 0.21 s_{32} + 0.17 s_{42} = 0.62 + 0.62 s_{22}$$

$$s_{22} = 1.63$$

$$s_{12} = 0.13 + 0.69 s_{12} + 0.13(1.63)$$

$$s_{12} = 1.10$$

$$s_{13} = 0.69 s_{13} + 0.13 s_{23} + 0.18 s_{43} = 0.69 s_{13} + 0.13 s_{23}$$

$$s_{23} = 0.21 + 0.62 s_{23} + 0.21 s_{33} + 0.17 s_{43} = 0.21 + 0.62 s_{23} + 0.21 s_{33}$$

$$s_{33} = 0.91 + 0.91 s_{33} + 0.09 s_{43} = 0.91 + 0.91 s_{33}$$

$$s_{33} = 10.1$$

$$s_{23} = 0.21 + 0.62 s_{23} + 0.21(10.1)$$

$$s_{23} = 6.13$$

$$s_{13} = 0.69 s_{13} + 0.13(6.13)$$

$$s_{13} = 2.57$$

$$\mu_1 = \alpha_1 + \alpha_1 s_{11} = 0.5 + 0.5(2.23) = 1.62$$

$$\mu_2 = \alpha_2 + \alpha_1 s_{12} + \alpha_2 s_{22} = 0.3 + 0.5(1.10) + 0.3(1.63) = 1.34$$

$$\mu_3 = \alpha_3 + \alpha_1 s_{13} + \alpha_2 s_{23} + \alpha_3 s_{33} = 0.2 + 0.5(2.57) + 0.3(6.13) + 0.2(10.1) = 5.34 \quad \square$$

Homework Problems

6.1 Introduction

(1) A basketball player attempts a sequence of shots from the free throw line. If she makes a shot, then the probability that she makes the next shot is 0.88. If she misses a shot, then the probability that she misses the next shot is 0.55. Model the sequence of shots using a discrete-time Markov chain.

(2) Consider the basketball player from problem (1). Suppose the probability that she makes a shot is dependent on the result of her two most recent shots. If she made both, then the probability that she makes the next shot is 0.92. If she made one, then the probability that she makes the next shot is 0.75. If she missed both, then the probability that she makes the next shot is 0.4. Model the sequence of shots using a discrete-time Markov chain.

(3) A math professor uses pop quizzes to evaluate the performance of students in his class. If he gave a pop quiz during his last lecture, then the probability of a quiz in his next lecture is 0.3. If he did not give a pop quiz during his last lecture, then the probability of a quiz in his next lecture is 0.8. Model the sequence of lectures (quiz or no quiz) using a discrete-time Markov chain. Note: This problem is adapted from Ross (2000).

(4) Consider the math professor from problem (3). Suppose the probability that he gives a quiz is dependent on the content of his two most recent lectures. If he gave a quiz in both, then the probability that he gives a quiz in the next lecture is 0.1. If he gave one quiz, then the probability that he gives a quiz in the next lecture is 0.4. If he gave no quizzes, then the probability that he gives a quiz in the next lecture is 0.9. Model the sequence of lectures (quiz or no quiz) using a discrete-time Markov chain.

(5) In a certain academic department, a faculty member can be one of an assistant professor, an associate professor, or a full professor. In terms of departmental standing, a full professor is above an associate professor, who is, in turn, above an assistant professor. At the end of each academic year, 13% of assistant professors become associate professors and 18% leave the department; 21% of associate professors become full professors and 17% leave the department; and 9% of full professors leave the department. Demotions never occur. Model the faculty career path using a discrete-time Markov chain.

(6) An electronic device is always in one of four states: perfect, degraded, poor, or failed. If a device is perfect at the beginning of one day, then at the beginning of the next day, there is a 90% chance that it is still perfect, a 6% chance that it is degraded, a 3% chance that it is poor, and a 1% chance that it is failed. If a device is degraded at the beginning of one day, then at the beginning of the next day, there is an 85% chance that it is still degraded, an 11% chance that it is poor, and a 4% chance that it is failed. If a device is poor at the beginning of one day, then at the beginning of the next day, there is an 80% chance that it is still poor, and a 20% chance that it is failed. Once a device is failed, it can never be restored to any other condition. Model the condition of the device using a discrete-time Markov chain.

(7) A diagnostic test is applied to determine the status of a component that is known to be in one of two states (state 1 and state 2). Let p denote the probability that the test indicates that the component is in state 1. The diagnostic test is applied repeatedly until a stopping condition is satisfied. Let $Y_1(t)$ denote the number of the first t applications of the test that indicate that the component is in state 1, and let $Y_2(t)$ denote the number of the first t applications of the test that indicate that the component is in state 2, $t = 1, 2, \ldots$. The applications of the diagnostic test terminate when

$$\left|Y_1(t) - Y_2(t)\right| = k$$

where k is some positive integer. Model the application of the diagnostic test using a discrete-time Markov chain.

(8) The maintenance supervisor of an assembly line has two tool cabinets, one at each end of the assembly line. Each morning, she walks from one end of the line to the other, and she is equally likely to begin the walk at either end. In the two tool cabinets are a total of six flashlights. At the beginning of her walk, the supervisor takes a flashlight (if one is available) from the tool cabinet at that location, and at the end of her walk, she leaves a flashlight (if she possesses one) from the tool cabinet at that location. Model the movement of flashlights using a discrete-time Markov chain.

(9) A maintenance technician for an assembly line stores his three toolkits at each end of the assembly line. Throughout the day, he walks back and forth, up and down the assembly line. At each end of the assembly line, he checks a computer monitor to determine if any maintenance requests have occurred. If there is a need for maintenance and a toolkit is available at his location, then he takes a toolkit with him for his next walk and takes care of the maintenance needs

along the way. He then leaves the toolkit at the end of his walk. If there is a need for maintenance and no toolkit is available at his location, then he must take care of the maintenance need on his next walk. If there is no need for maintenance, then he does not take a toolkit with him on his walk. The probability that one or more maintenance requests occur during a walk is p. Model the movement of toolkits and satisfaction of maintenance requests using a discrete-time Markov chain.

6.2 Manipulating the Transition Probability Matrix

(1) A machine's status is checked at the end of every hour and can be classified in one of three ways: 1, operating properly; 2, operating in a degraded condition; 3, failed. The machine's status at the end of an hour is only dependent on the status of the machine at the beginning of the hour. Thus, if we let $X(t)$ denote the status of the machine (1, 2, or 3) at the end of hour t, then $\{X(t), t = 0, 1, \ldots\}$ is a discrete-time Markov chain having transition probability matrix

$$\mathbf{P} = \begin{bmatrix} 0.7 & 0.2 & 0.1 \\ 0 & 0.6 & 0.4 \\ 0.6 & 0.3 & 0.1 \end{bmatrix}$$

(a) Given that the machine is observed to be operating properly at the end of one particular hour, what is the probability that it is observed to be failed two hours later?

(b) Given that the machine is observed to be operating properly at the end of one particular hour, what is the probability that it is observed to be failed three hours later?

(c) Given that the machine is observed to be operating properly at the end of one particular hour, what is the probability that it is observed to be failed four hours later?

(d) Suppose the machine is initially operating properly with probability 0.8 and degraded with probability 0.2. Determine the probabilities that the machine is in each of the three conditions after three hours.

(e) Given that the machine is observed to be operating properly at the end of one particular hour, what is the probability that it is first observed to be failed two hours later?

(f) Given that the machine is observed to be operating properly at the end of one particular hour, what is the probability that it is first observed to be failed three hours later?

(g) Given that the machine is observed to be operating properly at the end of one particular hour, what is the probability that it is first observed to be failed four hours later?

(h) Given that the machine is observed to be operating properly at the end of one particular hour, what is the expected value of the number of hours until it is first observed to be operated in a degraded condition? What is the expected value of the number of hours until it is first observed to be failed?

(2) The status of a popular book in a library is checked at the end of every week and can be classified in one of three ways: 1, on the shelf; 2, out but not overdue; 3, out and overdue. The book's status at the end of a week is only dependent on the status of the book at the beginning of the week. Thus, if we let $X(t)$ denote the status of the book (1, 2, or 3) at the end of week t, then $\{X(t), t = 0, 1, \ldots\}$ is a discrete-time Markov chain having transition probability matrix

$$\mathbf{P} = \begin{bmatrix} 0.3 & 0.7 & 0 \\ 0.2 & 0.5 & 0.3 \\ 0.2 & 0.1 & 0.7 \end{bmatrix}$$

(a) Given that the book is on the shelf at the end of one particular week, what is the probability that it is out but not overdue four weeks later?

(b) Suppose the book is initially on the shelf with probability 0.6, initially out but not overdue with probability 0.3, and initially out and overdue with probability 0.1. Determine the probabilities that the book is in each of the three classifications after three weeks.

(c) Given that the book is out and overdue at the end of one particular week, what is the probability that it is first observed to be out but not overdue two weeks later? Three weeks later? Four weeks later?

(d) Given that the book is out and overdue at the end of one particular week, what is the expected value of the number of weeks until it is first on the shelf?

(e) Given that the book is out and overdue at the end of one particular week, what is the expected value of the number of weeks until it is first out and overdue?

6.3 Classification of States

(1) Identify the classes of the Markov chain formulated in homework problem (1) of section 6.1 and specify if each class is recurrent or transient.

(2) Identify the classes of the Markov chain formulated in homework problem (2) of section 6.1 and specify if each class is recurrent or transient.

(3) Identify the classes of the Markov chain formulated in homework problem (3) of section 6.1 and specify if each class is recurrent or transient.

(4) Identify the classes of the Markov chain formulated in homework problem (4) of section 6.1 and specify if each class is recurrent or transient.

(5) Identify the classes of the Markov chain formulated in homework problem (5) of section 6.1 and specify if each class is recurrent or transient.

(6) Identify the classes of the Markov chain formulated in homework problem (6) of section 6.1 and specify if each class is recurrent or transient.

(7) Identify the classes of the Markov chain formulated in homework problem (7) of section 6.1 (assuming $k = 2$) and specify if each class is recurrent or transient.

(8) Identify the classes of the Markov chain formulated in homework problem (8) of section 6.1 and specify if each class is recurrent or transient.

(9) Identify the classes of the Markov chain formulated in homework problem (9) of section 6.1 and specify if each class is recurrent or transient.

6.4 Limiting Behavior

(1) Consider the discrete-time Markov chain formulated in homework problem (1) of section 6.1. Determine the player's long-term percentage of shots made.

(2) Consider the discrete-time Markov chain formulated in homework problem (2) of section 6.1.

(a) Determine the limiting probabilities for the Markov chain.
(b) Determine the player's long-term percentage of shots made.
(c) Given that the player missed her last shot, what is the probability that she makes her next shot?

(3) Consider the discrete-time Markov chain formulated in homework problem (3) of section 6.1. Determine the long-term probability that a quiz is given during a lecture.

(4) Consider the discrete-time Markov chain formulated in homework problem (4) of section 6.1.
(a) Determine the limiting probabilities for the Markov chain.
(b) Determine the long-term probability that a quiz is given during a lecture.
(c) Given that a quiz was given during the last lecture, what is the probability that a quiz is not given during the next lecture?

(5) Consider the discrete-time Markov chain formulated in homework problem (8) of section 6.1.
(a) Determine the limiting probabilities for the Markov chain.
(b) Determine the long-term probability that the supervisor walks without a flashlight.

(6) Consider the discrete-time Markov chain formulated in homework problem (9) of section 6.1 (assuming $p = 0.2$).
(a) Determine the limiting probabilities for the Markov chain.
(b) The maintenance department charges the production department \$50 for every walk made by the technician that involves maintenance. The production department charges the maintenance department \$20 for every walk made by the technician without a toolkit that goes past a maintenance need. Determine the long-run rate of funds transfers between the maintenance and production departments.

6.5 Absorbing States

(1) Consider the discrete-time Markov chain formulated in homework problem (5) of section 6.1. Suppose 60% of faculty members enter the department as an assistant professor, 30% as an associate professor, and 10% as a full professor. When an assistant professor leaves the department, 60% of the time it is on "bad terms" and 40% of the time it is on "good terms." When an associate professor leaves the department, 75% of the time it is on good terms. When a full professor leaves the department, 90% of the time it is on good terms.

(a) Modify the definition of $X(t)$ and the transition probability matrix to reflect this additional information.

(b) What is the probability that a faculty member leaves the department on bad terms?

(c) On average, how many years does a faculty member stay in the department?

(d) On average, how many years does a faculty member spend in each academic rank?

(2) Consider the discrete-time Markov chain formulated in homework problem (6) of section 6.1. All devices are purchased in the perfect state, and the state of the device is observed at the end of each day.

(a) On average, how many days elapse before the device is observed to be failed?

(b) On average, how many times is the device observed to be in each of the other three conditions?

(3) Consider the discrete-time Markov chain formulated in homework problem (7) of section 6.1 (assuming $k = 2$ and $p = 0.8$).

(a) What is the probability that the sequence of tests terminates with more individual test results indicating that the component is in state 1?

(b) On average, how many individual tests will be conducted?

Application: Inventory Management

A local store uses an (s, S) inventory policy for a particular product. Every Friday evening after the store closes, the inventory level is checked. If the stock level is greater than s, then no action is taken. Otherwise, enough stock is procured over the weekend so that, when the store reopens on Monday morning, the inventory level is $S = 8$. Suppose the demand for the product during a given week is a Poisson random variable having a mean of 4 items per week. If the demand exceeds the inventory during a given week, that is, if a shortage occurs, then the excess demand is lost.

Task 1: Let $f(n)$ denote the probability that the demand is n during a given week, and let $R(n)$ denote the probability that the demand is greater than n during a given week. Construct expressions for $f(n)$ and $R(n)$.

Task 2: Define $X(t)$ so that $\{X(t), t = 0, 1, \dots\}$ can be modeled as a discrete-time Markov chain.

The store pays \$5 in inventory costs for every unit on the shelves at the end of the week. In addition, the store estimates it loses \$35 in future sales when it has a week with a shortage. Replenishment orders cost the store \$15 plus \$10 per item. The store sells each item for \$22. Suppose that $s = 3$.

Task 3: Construct the transition probability matrix for $\{X(t), t = 0, 1, \dots\}$ using $f(n)$ and $R(n)$.

Task 4: If the starting inventory on Monday morning is 6 units, determine the probability that a shortage occurs by Friday afternoon.

Task 5: If the starting inventory on Monday morning is 6 units, determine the expected value of the inventory level on Friday afternoon.

Task 6: If the starting inventory on Monday morning is 6 units, determine the probability that the store orders k items of replenishment stock on Friday afternoon (for all possible values of k).

Task 7: If the starting inventory on Monday morning is 6 units, determine the probability that the store sells k items by Friday afternoon (for all possible values of k).

Task 8: Determine the limiting probabilities for the Markov chain.

Task 9: Determine the long-run probability that a shortage occurs during a given week.

Task 10: Determine the long-run probability of ordering k items at the end of a given week (for all possible values of k).

Task 11: Determine the long-run probability of selling k items during a given week (for all possible values of k).

Task 12: Determine the store's long-run average profit per week.

Task 13: What value of s should the store use?

7

Continuous-Time Markov Chains

In this chapter, we consider continuous-time Markov chains, the class of discrete-valued, continuous-time stochastic processes that possess the Markov property. Continuous-time Markov chains are used in a wide variety of applications, including machine repair and queueing problems.

7.1 Introduction

The definition of a continuous-time Markov chain is a natural extension of the corresponding definition of discrete-time Markov chains.

> A discrete-valued, stochastic process $\{X(t),\ t \ge 0\}$ is a **continuous-time Markov chain** if and only if
>
> $$\Pr\left(X(t+s)=j \mid X(s)=i;\ X(u)=x(u),\ 0\le u<s\right)=\Pr\left(X(t+s)=j \mid X(s)=i\right)$$
>
> for all $s > 0$, $t \ge 0$ and for all i, j, and $x(u)$ in the state space. Furthermore,
>
> $$P_{ij}(t)=\Pr\left(X(t+s)=j \mid X(s)=i\right)$$
>
> is referred to as the **transition probability function** from state i to state j in time t.

The fact that a continuous-time Markov chain has stationary transition probabilities leads to a second definition of a continuous-time Markov chain.

> A **continuous-time Markov chain** is a discrete-valued, continuous-time stochastic process having the properties that each time it enters any state i in the state space:
>
> 1. The amount of time it spends in that state before making a transition to a different state (T_i) is an exponential random variable with rate v_i.
> 2. When the process leaves state i, it next enters another state in the state space j with probability P_{ij}.

Let q_{ij} denote the **transition rate** from state i to state j where

$$q_{ij} = P_{ij}\nu_i$$

Under this definition, $P_{ii} = 0$ and

$$\sum_j q_{ij} = \nu_i$$

A reasonable way to think of a continuous-time Markov chain is that it has transition rules that are the same as a discrete-time Markov chain, but the residence times in the states before transitions are exponential random variables.

The easiest way to model a system as a continuous-time Markov chain is to define the state of the system $X(t)$ and either specify all positive transition rates or draw a rate diagram. In a rate diagram, each state is represented by a node. The nodes are connected with directed arcs that represent potential transitions, and each arc is labeled with the associated transition rate.

Example 7.1

A facility contains four independent and identical machines that are serviced by two equally trained technicians. The time to failure for a machine is an exponential random variable having rate 0.05 failure per hour, and the time required for one technician to repair a machine is an exponential random variable having rate 0.25 repair per hour.

(a) Model this scenario using a continuous-time Markov chain.

Let $X(t)$ denote number of machines operating at time t, $t \geq 0$.

The rate diagram is given below. The transition rate from state 4 to state 3 is 0.2 due to the fact that four machines are operating with each machine having a failure rate of 0.05. The transition rate from state 0 to state 1 is 0.5 because two technicians are each working at a rate of 0.25.

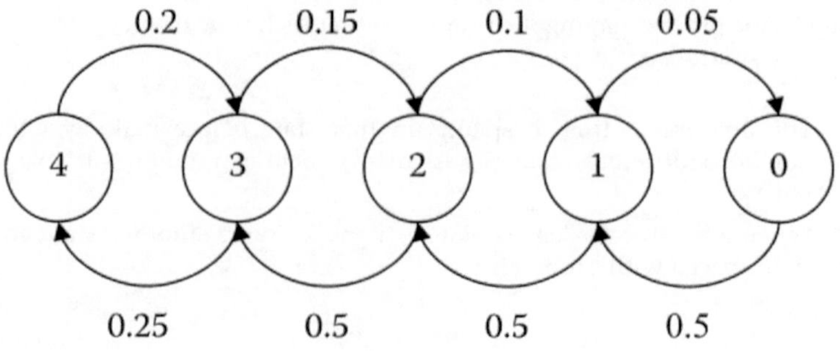

(b) Determine the transition rates for each state.

$$\nu_0 = q_{01} = 0.5$$

$$\nu_1 = q_{10} + q_{12} = 0.05 + 0.5 = 0.55$$

$$\nu_2 = q_{21} + q_{23} = 0.1 + 0.5 = 0.6$$

$$\nu_3 = q_{32} + q_{34} = 0.15 + 0.25 = 0.4$$

$$\nu_4 = q_{43} = 0.2$$

(c) Determine the transition probabilities for the process.

$$P_{01} = 1$$

$$P_{10} = \frac{0.05}{0.55} = \frac{1}{11} \qquad P_{12} = \frac{10}{11}$$

$$P_{21} = \frac{0.1}{0.6} = \frac{1}{6} \qquad P_{23} = \frac{5}{6}$$

$$P_{32} = \frac{0.15}{0.4} = \frac{3}{8} \qquad P_{34} = \frac{5}{8}$$

$$P_{43} = 1 \square$$

Example 7.2

Consider an automated machine that processes parts in two steps, step 1 and step 2. Each step occurs at a separate station within the machine; step 1 occurs at station 1, and step 2 occurs at station 2. Upon arrival, a part completes step 1, and then moves on to step 2. Parts arrive according to a Poisson process having rate λ, but parts are only accepted by the machine if both stations are empty. The time required to complete step i is an exponential random variable having rate μ_i, $i = 1, 2$.

(a) Model this scenario using a continuous-time Markov chain.

$$\text{Let } X(t) = \begin{cases} 0 & \text{if both stations are empty at time } t \\ 1 & \text{if station 1 is occupied at time } t \\ 2 & \text{if station 2 is occupied at time } t \end{cases} \qquad t \geq 0$$

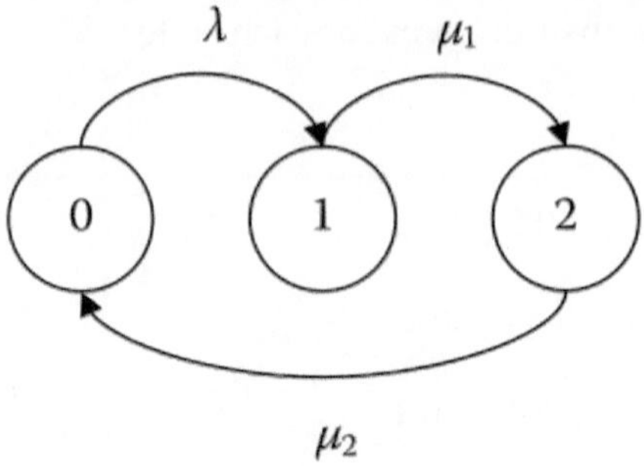

(b) Determine the transition rates for each state.

$\nu_0 = \lambda$

$\nu_1 = \mu_1$

$\nu_2 = \mu_2$

(c) Determine the transition probabilities for the process.

$P_{01} = P_{12} = P_{20} = 1$ □

7.2 Birth and Death Processes

Suppose entities enter and leave a system one at a time. Suppose that whenever there are n entities in the system: (a) new entities enter the system (are "born") at an exponential rate λ_n, and (b) entities leave the system ("die") at an exponential rate μ_n (except when $n = 0$). So, whenever there are n entities in the system, the time until the next arrival (or "birth") is an exponential random variable having rate λ_n, and the time until the next departure (or "death") is an exponential random variable having rate μ_n (except when $n = 0$). Furthermore, arrivals and departures are independent occurrences. Such a system can be modeled as a continuous-time Markov chain, where the state of the system is the number of entities (people) in the system at that time. Such a process is called a birth and death process.

A **birth and death process** is a continuous-time Markov chain $\{X(t), t \geq 0\}$ with state space $\{0, 1, \ldots\}$ such that

1. From state 0, the process always transitions to state 1.
2. From all other states $n = 1, 2, \ldots$, the process always transitions to either state $n - 1$ or state $n + 1$.

For such a process,

$$q_{n,n+1} = \lambda_n$$

and

$$q_{n,n-1} = \mu_n$$

$n = 1, 2, \ldots$. In addition,

$$\upsilon_0 = \lambda_0$$

$$\upsilon_n = \lambda_n + \mu_n$$

$n = 1, 2, \ldots$,

$$P_{01} = 1$$

$$P_{n,n+1} = \frac{\lambda_n}{\lambda_n + \mu_n}$$

$n = 1, 2, \ldots$, and

$$P_{n,n-1} = \frac{\mu_n}{\lambda_n + \mu_n}$$

$n = 1, 2, \ldots$.

The sequence $\{\lambda_0, \lambda_1, \ldots\}$ is referred to as the sequence of **birth rates** for the process. The sequence $\{\mu_1, \mu_2, \ldots\}$ is referred to as the sequence of **death rates** for the process. If $\mu_n = 0$ for all $n = 1, 2, \ldots$, then $\{X(t), t \geq 0\}$ is referred to as a **pure birth process.**

Example 7.3

Model the Poisson process as a pure birth process.

Let $X(t)$ denote the number of arrivals by time t, $t \geq 0$.

$\lambda_n = \lambda$, $n = 0, 1, \ldots$

$\mu_n = 0$, $n = 1, 2, \ldots$ □

Example 7.4

Consider the following queueing system. Customers arrive according to a Poisson process having rate λ. Upon arrival, the customers wait in a single, first-in first-out queue (waiting line) for a single server. The time required to serve a customer is an exponential random variable having rate μ. This system is referred to as the M/M/1 queue. Model the M/M/1 queue as a birth and death process.

Let $X(t)$ denote the number of customers in the system at time t, $t \geq 0$.

$$\lambda_n = \lambda,\ n = 0, 1, \ldots$$

$$\mu_n = \mu,\ n = 1, 2, \ldots \square$$

Example 7.5

Consider the following queueing system. Customers arrive according to a Poisson process having rate λ. Upon arrival, the customers wait in a single, first-in first-out queue for one of s identical servers. The time required to serve a customer is an exponential random variable having rate μ. This system is referred to as the M/M/s queue. Model the M/M/3 queue as a birth and death process.

Let $X(t)$ denote the number of customers in the system at time t, $t \geq 0$.

$$\lambda_n = \lambda,\ n = 0, 1, \ldots$$

$$\mu_n = n\mu,\ n = 1, 2, 3$$

$$\mu_n = 3\mu,\ n = 4, 5, \ldots \square$$

7.3 Limiting Probabilities

When we considered discrete-time Markov chains, we analyzed the behavior of the process in two ways: transient behavior (such as n-step transition probabilities, n-step first passage times, n-step occupancy, etc.) and steady-state behavior (limiting/stationary probabilities). We continue our study of continuous-time Markov chains with steady-state behavior.

Let $\{X(t), t \geq 0\}$ be a continuous-time Markov chain such that the matrix of transition probabilities corresponds to the transition probability matrix of an irreducible discrete-time Markov chain. If ρ_i denotes the **limiting probability** of state i for all i in the state space, then

$$\rho_i = \lim_{t \to \infty} \Pr(X(t) = i)$$

If we can compute the limiting probabilities for the continuous-time Markov chain, then we can answer many important questions about the long-term behavior of the process.

The method used to compute the limiting probabilities is based on the concept of steady-state behavior and the construction of balance equations. At "steady state," the total rate of transition into a state must equal the total rate of transition out of the state. We find the limiting probabilities for a continuous-time Markov chain by constructing the balance equations ("rate in" equals "rate out") for all but one state and requiring that the limiting probabilities sum to one.

Example 7.1 (continued)

(a) Determine the limiting probabilities for the process.

State 0

total rate in = $0.05\ \rho_1$

total rate out = $0.5\ \rho_0$

$0.05\ \rho_1 = 0.5\ \rho_0$

$\rho_1 = 10\ \rho_0$

State 1

$0.5\ \rho_0 + 0.1\ \rho_2 = 0.55\ \rho_1$

$0.5\ \rho_0 + 0.1\ \rho_2 = 5.5\ \rho_0$

$\rho_2 = 50\ \rho_0$

State 2

$0.5\ \rho_1 + 0.15\ \rho_3 = 0.6\ \rho_2$

$5\ \rho_0 + 0.15\ \rho_3 = 30\ \rho_0$

$\rho_3 = (500/3)\ \rho_0$

State 4

$0.25\ \rho_3 = 0.2\ \rho_4$

$(125\ /\ 3)\ \rho_0 = 0.2\ \rho_4$

$\rho_4 = (625/3)\ \rho_0$

Sum

$\rho_0 + \rho_1 + \rho_2 + \rho_3 + \rho_4 = 1$

$\rho_0 + 10\ \rho_0 + 50\ \rho_0 + (500/3)\ \rho_0 + (625/3)\ \rho_0 = 1$

$(1308\ /\ 3)\ \rho_0 = 1$

$\rho_0 = 0.0023$ $\quad \rho_1 = 0.0229$ $\quad \rho_2 = 0.1147$ $\quad \rho_3 = 0.3823$ $\quad \rho_4 = 0.4778$

(b) In the long-run, on average, how many machines are operating?

$$0\ \rho_0 + 1\ \rho_1 + 2\ \rho_2 + 3\ \rho_3 + 4\ \rho_4 = 3.31$$

(c) At steady state, on average, how many operators are busy?

$$2\ \rho_0 + 2\ \rho_1 + 2\ \rho_2 + 1\ \rho_3 + 0\ \rho_4 = 0.66$$

(d) What is the long-run proportion of time that both operators are busy?

$$\rho_0 + \rho_1 + \rho_2 = 0.1399$$

(e) Suppose each hour of machine downtime costs \$200 and operators are paid \$25 per hour. At steady state, on average, how much money is being "wasted" per 8-hour day?

$$8(\$200)(4 - 3.31) + 8(\$25)(2 - 0.66) = \$1372 \square$$

Example 7.4 (continued)

(a) Determine the limiting probabilities for the process.

State 0
$$\mu\rho_1 = \lambda\rho_0$$
$$\rho_1 = \frac{\lambda}{\mu}\rho_0$$

State 1
$$\lambda\rho_0 + \mu\rho_2 = (\lambda+\mu)\rho_1$$
$$\lambda\rho_0 + \mu\rho_2 = (\lambda+\mu)\frac{\lambda}{\mu}\rho_0$$
$$\mu\rho_2 = (\lambda+\mu)\frac{\lambda}{\mu}\rho_0 - \lambda\rho_0 = \left(\frac{\lambda^2}{\mu} + \lambda - \lambda\right)\rho_0$$
$$\rho_2 = \frac{\lambda^2}{\mu^2}\rho_0 = \left(\frac{\lambda}{\mu}\right)^2\rho_0$$

Claim: $\rho_n = \left(\frac{\lambda}{\mu}\right)^n \rho_0 \quad n = 1, 2, \ldots$

Proof: by induction

$$\lambda\rho_{k-1} + \mu\rho_{k+1} = (\lambda+\mu)\rho_k$$
$$\lambda\left(\frac{\lambda}{\mu}\right)^{k-1}\rho_0 + \mu\rho_{k+1} = (\lambda+\mu)\left(\frac{\lambda}{\mu}\right)^k \rho_0$$
$$\mu\rho_{k+1} = (\lambda+\mu)\left(\frac{\lambda}{\mu}\right)^k \rho_0 - \lambda\left(\frac{\lambda}{\mu}\right)^{k-1}\rho_0 = \left(\frac{\lambda^{k+1}}{\mu^k} + \frac{\lambda^k}{\mu^{k-1}} - \frac{\lambda}{\mu^{k-1}}\right)\rho_0$$
$$\rho_{k+1} = \left(\frac{\lambda}{\mu}\right)^{k+1}\rho_0$$

Note $\rho_n = \left(\frac{\lambda}{\mu}\right)^n \rho_0$ for $n = 0, 1, \ldots$.

Sum $\sum_{n=0}^{\infty} \rho_n = \sum_{n=0}^{\infty} \left(\frac{\lambda}{\mu}\right)^n \rho_0 = \rho_0 \sum_{n=0}^{\infty} \left(\frac{\lambda}{\mu}\right)^n = \rho_0 \frac{1}{1-\frac{\lambda}{\mu}} = 1$ if $\lambda < \mu$

Thus, $\rho_0 = \left(1 - \frac{\lambda}{\mu}\right)$

$$\rho_n = \left(\frac{\lambda}{\mu}\right)^n \left(1 - \frac{\lambda}{\mu}\right) \quad n = 0, 1, \ldots$$

(b) What is the long-run average number of customers in the system?

$$\sum_{n=0}^{\infty} n\rho_n = \sum_{n=0}^{\infty} n \left(\frac{\lambda}{\mu}\right)^n \left(1 - \frac{\lambda}{\mu}\right) = \left(1 - \frac{\lambda}{\mu}\right) \sum_{n=0}^{\infty} n \left(\frac{\lambda}{\mu}\right)^n = \left(1 - \frac{\lambda}{\mu}\right) \frac{\lambda/\mu}{(1-\lambda/\mu)^2}$$

$$\frac{\lambda/\mu}{(1-\lambda/\mu)}$$

(c) What is the long-run proportion of the time that the server is idle?

$$\rho_0 = 1 - \frac{\lambda}{\mu} \quad \square$$

7.4 Time-Dependent Behavior

The limiting probabilities for a continuous-time Markov chain capture the steady-state behavior of the system under consideration. But, what if we are interested in system behavior over some shorter period of time? In this case, we analyze the behavior of the system using the time-dependent transition probabilities.

If $\{X(t), t \geq 0\}$ is a continuous-time Markov chain having state space K, then, as given by the Kolmogorov backward differential equations,

$$\frac{dP_{ij}(t)}{dt} = \sum_{k \neq i} q_{ik} P_{kj}(t) - v_i P_{ij}(t)$$

for all $t \geq 0$, and for all i and j in the state space.

Example 7.6

A single machine operates for an amount of time that is exponentially distributed with rate λ. After a failure, the machine is restored to an operating condition (repaired) in an amount of time that is exponentially distributed with rate μ. When evaluating the performance of such equipment, it is common to compute time-dependent availability, $A(t)$, the probability that the machine is operating at time t. Construct an expression for $A(t)$.

$$X(t)=\begin{cases}1 & \text{if the machine is operating at time } t\\ 0 & \text{if the machine is failed at time } t\end{cases}$$

$$P_{01} = P_{10} = 1$$

$$\nu_0 = \mu$$

$$\nu_1 = \lambda$$

$$\frac{dP_{11}(t)}{dt} = q_{10}P_{01}(t) - \nu_1 P_{11}(t) = \lambda P_{01}(t) - \lambda P_{11}(t) \tag{7.1}$$

$$\frac{dP_{01}(t)}{dt} = q_{01}P_{11}(t) - \nu_0 P_{01}(t) = \mu P_{11}(t) - \mu P_{01}(t) \tag{7.2}$$

Multiply (7.1) by μ and (7.2) by λ.

$$\mu\frac{dP_{11}(t)}{dt} = \lambda\mu P_{01}(t) - \lambda\mu P_{11}(t) \tag{7.3}$$

$$\lambda\frac{dP_{01}(t)}{dt} = \lambda\mu P_{11}(t) - \lambda\mu P_{01}(t) \tag{7.4}$$

Add (7.3) to (7.4).

$$\mu\frac{dP_{11}(t)}{dt} + \lambda\frac{dP_{01}(t)}{dt} = 0 \tag{7.5}$$

Integrate (7.5).

$$\mu P_{11}(t) + \lambda P_{01}(t) = C \tag{7.6}$$

Since $P_{11}(0) = 1$ and $P_{01}(0) = 0$, $C = \mu$.

$$\mu P_{11}(t) + \lambda P_{01}(t) = \mu \tag{7.7}$$

$$\lambda P_{01}(t) = \mu\left[1 - P_{11}(t)\right] \tag{7.8}$$

Substitute (7.8) into (7.1).

$$\frac{dP_{11}(t)}{dt} = \mu\left[1 - P_{11}(t)\right] - \lambda P_{11}(t) = \mu - (\lambda + \mu)P_{11}(t) \tag{7.9}$$

Let

$$h(t) = P_{11}(t) - \frac{\mu}{\lambda + \mu} \tag{7.10}$$

$$\frac{dh(t)}{dt} = \frac{dP_{11}(t)}{dt} \tag{7.11}$$

Rewrite (7.9) using (7.10) and (7.11).

$$\frac{dh(t)}{dt} = \mu - (\lambda + \mu)\left[h(t) + \frac{\mu}{\lambda + \mu}\right] = -(\lambda + \mu)h(t)$$

$$\frac{dh(t)/dt}{h(t)} = -(\lambda + \mu) \tag{7.12}$$

Integrate both sides of (7.12).

$$\ln\left[h(t)\right] = -(\lambda + \mu)t + C$$

$$h(t) = Ke^{-(\lambda + \mu)t} \tag{7.13}$$

Substitute (7.13) into (7.10).

$$P_{11}(t) = \frac{\mu}{\lambda + \mu} + Ke^{-(\lambda + \mu)t} \tag{7.14}$$

Recall that $P_{11}(0) = 1$.

$$P_{11}(0) = \frac{\mu}{\lambda + \mu} + K = 1$$

$$K = \frac{\lambda}{\lambda + \mu} \tag{7.15}$$

Substitute (7.15) into (7.14).

$$P_{11}(t) = A(t) = \frac{\mu}{\lambda + \mu} + \frac{\lambda}{\lambda + \mu} e^{-(\lambda+\mu)t} \quad \square$$

The two-state process considered in Example 7.6 is the simplest possible continuous-time Markov chain, but in spite of that simplicity, analytic evaluation of the time-dependent transition probability functions is quite difficult. For this reason, discrete-event simulation is often used to estimate the time-dependent behavior of such processes.

7.5 Semi-Markov Processes

What if a stochastic process possesses all the properties of a continuous-time Markov chain with the exception that the residence times in states are not exponentially distributed? Such processes are referred to as semi-Markov processes.

Let $\{X(t), t \geq 0\}$ be a continuous-time stochastic process having state space $\{1, 2, \ldots, N\}$. Suppose that once the process enters state i, the amount of time until it exits state i is a random variable having mean μ_i, $i = 1, 2, \ldots, N$. If the transitions from state to state are governed by the transition probability matrix of an irreducible discrete-time Markov chain, then $\{X(t), t \geq 0\}$ is said to be a **semi-Markov process**. If π denotes the limiting probability vector of the underlying discrete-time Markov chain and

$$\theta_i = \lim_{t \to \infty} \Pr(X(t) = i)$$

$i = 1, 2, \ldots, N$, then

$$\theta_i = \frac{\pi_i \mu_i}{\sum_{j=1}^{N} \pi_j \mu_j}$$

$i = 1, 2, \ldots, N$.

Homework Problems

7.2 Birth and Death Processes

(1) Consider the following machine repair problem. A facility contains four identical machines that operate independently. Once a machine starts operating, the time until it fails is an exponential random variable having rate $\lambda = 0.1$ failure per hour. The time required for a technician to restore a failed machine to its original status is an exponential random variable having rate $\mu = 0.4$ repair per hour. Three equally trained technicians are assigned the duty of repairing failed machines. Describe how this process can be modeled using a continuous-time Markov chain. Include in your description: (a) the definition of $X(t)$, (b) a rate diagram, (c) the v_is, and (d) the P_{ij}s.

(2) Consider the following scenario. Customers arrive at a barber shop according to a Poisson process having rate $\lambda = 6$ customers per hour. The barber shop employs two barbers. The time required for a barber to provide services for a customer is an exponential random variable having rate $\mu = 4$ customers per hour. The barber shop only has three chairs for waiting customers. When a customer arrives and finds both barbers working and all three chairs occupied, the customer immediately leaves. Describe how this process can be modeled using a continuous-time Markov chain. Include in your description: (a) the definition of $X(t)$, (b) a rate diagram, (c) the v_is, and (d) the P_{ij}s.

(3) Consider the following machine repair problem. A facility contains two machines that operate independently. Once machine 1 starts operating, the time until it fails is an exponential random variable having rate $\lambda_1 = 0.1$ failure per hour. Once machine 2 starts operating, the time until it fails is an exponential random variable having rate $\lambda_2 = 0.2$ failure per hour. The time required for a technician to restore a failed machine 1 to its original status is an exponential random variable having rate $\mu_1 = 0.5$ repair per hour. The time required for a technician to restore a failed machine 2 to its original status is an exponential random variable having rate $\mu_2 = 0.4$ repair per hour. Two equally trained technicians are assigned the duty of repairing failed machines. Describe how this process can be modeled using a continuous-time Markov chain. Include in your description: (a) the definition of $X(t)$, (b) a rate diagram, (c) the v_is, and (d) the P_{ij}s.

7.3 Limiting Probabilities

(1) Consider the machine repair problem defined in problem (1) of section 7.2. Determine the following:

(a) the long-run average number of busy technicians

(b) the long-run average number of machines operating

(c) the long-run average number of machines waiting for repair

(2) Consider the machine repair problem defined in problem (1) of section 7.2. The manager of the facility is not convinced that three is the best choice for the number of technicians. Every hour of machine downtime costs the facility $275. Technicians are paid $30 per hour. How many technicians should the facility employ?

(3) Consider the barber shop problem defined in problem (2) of section 7.2. Determine the following:

(a) the long-run average number of customers being serviced

(b) the long-run average number of customers in the shop

(c) the long-run proportion of time both barbers are busy

(d) the long-run number of lost customers per hour

(4) Consider the barber shop problem defined in problem (2) of section 7.2. Suppose barbers are paid $6 per hour and customers are charged $12 each for services rendered by the barber. Would adding a third barber improve the profitability of the barber shop? Note that adding a third barber would require removal of one of the chairs for waiting customers.

(5) Consider the machine repair problem defined in problem (3) of section 7.2. Determine the following:

(a) the long-run average number of machines operating

(b) the long-run average number of busy technicians

(c) the long-run proportion of time machine 1 is operating

7.5 Semi-Markov Processes

(1) A small corporation employs a private pilot to take its executives on business trips. This pilot's contract is worded such that he is always in one of three states: off duty, stand-by, or active duty. The amount of time the pilot stays off duty is a random variable having a mean of 6 days, and 90% of off-duty periods end with a call to active duty. The amount of time the pilot stays on stand-by is a random variable having a mean of 3 days, and 70% of stand-by periods end with a call to active duty. The amount of time the pilot stays on active duty is a random variable having a mean of 14 days, and 60% of active-duty periods end with a downgrade to stand-by duty. The pilot is paid a wage of \$500 per day when he is on active duty and a wage of \$200 per day when he is on stand-by duty. The pilot is not paid when off duty. What is the pilot's long-run wage?

8

Markovian Queueing Systems

Queueing theory refers to the analysis of a class of stochastic processes that are used to model waiting lines, including the arrival of customers to servers, the waiting of customers for servers, the processing of customers by servers, and the departure of customers. In addition to queues containing human customers, the models presented in this chapter can be applied to queues of manufactured parts in an assembly process, calls in a telephone network, etc.

8.1 Queueing Basics

We begin our study of queueing systems by defining some fundamental terminology.

> Consider a queueing system. Let $T(n)$ denote the **interarrival time** of customer n, the elapsed time between the arrival of customer $n - 1$ and customer n, and let $S(n)$ denote the **service time** of customer n, $n = 1, 2, \ldots$. The stochastic process $\{T(n), n = 1, 2, \ldots\}$ is referred to as the **arrival process**, and the stochastic process $\{S(n), n = 1, 2, \ldots\}$ is referred to as the **service process**.

Until further notice, we assume that:

1. Customer $n = 0$ arrives at time 0
2. $\{T(1), T(2), \ldots\}$ is a sequence of independent and identically distributed random variables
3. $\{S(1), S(2), \ldots\}$ is a sequence of independent and identically distributed random variables
4. $\{T(1), T(2), \ldots\}$ and $\{S(1), S(2), \ldots\}$ are independent

Next, we characterize the average behavior of the arrival process.

> Consider a queueing system. If
>
> $$\lambda = 1/E[T(n)]$$
>
> then λ is referred to as the **arrival rate** for the system.

The arrival and service process only tell part of the story for a queueing system. We must also specify the number of servers, the number of waiting lines (one, one per server, etc.), the system capacity (if it is finite), the service discipline (first-in first-out or first-come first-served, last-in first-out, shortest processing time, etc.), and the size of the customer population. Throughout this chapter, we assume that there is only one waiting line and customers are serviced on a first-in first-out (FIFO) basis. We describe the remaining features of the queueing system using **Kendall's notation** (Kendall, 1953). Under Kendall's notation, a queueing system is represented by A/B/s/c/m where: A indicates the probability distribution of $T(n)$, B indicates the probability distribution of $S(n)$, s denotes the number of servers, c denotes the capacity of the system (customers who arrive when the system contains c customers **balk**, that is, they immediately depart without service), and m denotes the size of the customer population. If $m = \infty$, then the last portion of Kendall's notation is dropped. If $c = m = \infty$, then the last two portions of Kendall's notation are dropped.

A wide variety of queueing systems have been studied for many years. However, the most common choices for A include:

M	exponential (Markovian) interarrival times
E_j	j-Erlang interarrival times
D	deterministic (constant) interarrival times
G	no assumed interarrival time behavior

and the most common choices for B include:

M	exponential (Markovian) service times
E_j	j-Erlang service times
D	deterministic (constant) service times
G	no assumed service time behavior

Since the possibility exists that not all arriving customers will enter the system (some may balk), we need an additional definition.

Consider a queueing system. Let $N(t)$ denote the number of customers who enter the system in the interval $(0, t]$. If λ_a denotes the arrival rate of customers who actually enter the system, then λ_a is given by

$$\lambda_a = \lim_{t \to \infty} \frac{N(t)}{t}$$

As is the case in almost all analyses of queueing systems, we focus our attention on the long-run (limiting, stationary, steady-state) behavior of the system. There are nine commonly used, fundamental measures of the steady-state or long-run performance of a queueing system.

L	the expected value of the number of customers in the system
L_Q	the expected value of the number of customers waiting in queue
L_S	the expected value of the number of customers in service
W	the expected value of the amount of time a customer spends in the system
W_Q	the expected value of the amount of time a customer spends waiting in queue
W_S	the expected value of the amount of time a customer spends in service
p_n	the probability that there are n customers in the system
a_n	the probability that an arriving customer finds n customers in the system
d_n	the probability that a departing customer leaves n customers in the system

Several of these measures are related by the most famous result from queueing theory, **Little's Law** (Little, 1961).

> For a queueing system in which customers arrive, wait for and receive service, and then depart,
>
> $$L = \lambda_a W$$
>
> $$L_Q = \lambda_a W_Q$$
>
> and
>
> $$L_S = \lambda_a W_S$$

Note that the definitions of p_n, a_n, and d_n are quite similar. In some cases, a_n and d_n are equivalent.

> For an infinite-capacity queueing system in which customers arrive and are serviced individually,
>
> $$a_n = d_n$$
>
> $n = 0, 1, \ldots$.

At steady state, a queueing system is in equilibrium. For every customer who arrives and increases the number in the system from n to $n + 1$, there must be exactly one departure that causes the number in system to decrease from $n + 1$ to n. Otherwise, the number of customers in the system would either stay at zero or increase without bound (assuming there is no finite capacity on the system).

When the arrival process is Poisson, p_n and a_n are equivalent. This result is known as the **PASTA principle** (Poisson Arrivals See Time Averages).

For a queueing system in which customers arrive according to a Poisson process, that is, interarrival times are exponentially distributed,

$$p_n = a_n$$

$n = 0, 1, \ldots$.

8.2 The M/M/1 Queue

We now begin our study of specific types of queueing systems. Initially, we focus on queueing systems in which interarrival times and service times are exponentially distributed. These restrictive assumptions permit system analysis using continuous-time Markov chains. Therefore, these queueing systems are often referred to as **Markovian queueing systems**.

The first queueing system we consider is one that we considered in our study of continuous-time Markov chains—the M/M/1 queue. Interpretation of Kendall's notation indicates the following characteristics of M/M/1 queues: exponential interarrival times (rate λ), exponential service times (rate μ), one server, infinite capacity, and infinite customer population.

In our previous study of the M/M/1 queue, we found that

$$p_n = \left(\frac{\lambda}{\mu}\right)^n \left(1 - \frac{\lambda}{\mu}\right)$$

$n = 0, 1, \ldots$, and

$$L = \frac{\lambda}{\mu - \lambda}$$

Recall that these results required that $\lambda < \mu$. If the interarrival rate exceeds the service rate, then the server cannot keep up with customer arrivals, and the number of customers in the system would increase without bound (since there is no finite capacity).

These initial results allow the derivation of the seven remaining measures of long-run system performance.

Since no customers are denied entry,

$$\lambda_a = \lambda$$

By Little's Law,

$$W = L / \lambda = 1 / (\mu - \lambda)$$

Since $S(n) \sim \text{expon}(\mu)$,

$$W_S = 1 / \mu$$

Since $W = W_Q + W_S$

$$W_Q = W - W_S = \lambda / [\mu(\mu - \lambda)]$$

By Little's Law,

$$L_Q = \lambda W_Q = \lambda^2 / [\mu(\mu - \lambda)]$$

and

$$L_S = \lambda W_S = \lambda / \mu$$

By the PASTA principle,

$$a_n = p_n$$

Since the system has infinite capacity and customers arrive and are serviced individually,

$$d_n = a_n = p_n$$

Note that the **probability that the server is busy** is $1 - p_0 = \lambda/\mu$. For this reason, the quantity λ/μ is often denoted by ρ and referred to as the **utilization factor** for the system.

8.3 The M/M/1/c Queue

Next we consider the M/M/1/c queue. Interpretation of Kendall's notation indicates the following characteristics of M/M/1/c queues: exponential interarrival times (rate λ), exponential service times (rate μ), one server, finite system capacity of c, and infinite customer population. Customers who arrive when the system is full, that is, when there are c customers in the system, immediately depart without entering the system. These customers are said to have balked.

We can utilize a continuous-time Markov chain to construct expressions for p_n, $n = 0, 1, \dots, c$. Let $X(t)$ denote the number of customers in the system at time t. Note that $X(t) \in \{0, 1, \dots, c\}$ for all $t \geq 0$.

States $0, 1, \dots, c-1$ are identical to the M/M/1. Therefore,

$$p_n = \left(\frac{\lambda}{\mu}\right)^n p_0$$

$n = 0, 1, \dots, c$. Thus,

$$\sum_{n=0}^{c} p_n = \sum_{n=0}^{c} \left(\frac{\lambda}{\mu}\right)^n p_0 = p_0 \sum_{n=0}^{c} \left(\frac{\lambda}{\mu}\right)^n = p_0 \frac{1-(\lambda/\mu)^{c+1}}{1-(\lambda/\mu)} = 1$$

So,

$$p_n = \left(\frac{\lambda}{\mu}\right)^n \frac{1-(\lambda/\mu)}{1-(\lambda/\mu)^{c+1}}$$

$n = 0, 1, \dots, c$.

Note that these results did not require that $\lambda < \mu$. These results allow the derivation of the remaining measures of long-run system performance where $\rho = \lambda/\mu$.

We compute L directly.

$$L = \sum_{n=0}^{c} n p_n = \frac{1-\rho}{1-\rho^{c+1}} \sum_{n=0}^{c} n\rho^n = \frac{\rho(1-\rho)}{1-\rho^{c+1}} \sum_{n=0}^{c} n\rho^{n-1} = \frac{\rho(1-\rho)}{1-\rho^{c+1}} \sum_{n=0}^{c} \frac{d}{d\rho}\rho^n$$

$$L = \frac{\rho(1-\rho)}{1-\rho^{c+1}} \frac{d}{dp} \sum_{n=0}^{c} \rho^n = \frac{\rho(1-\rho)}{1-\rho^{c+1}} \frac{d}{dp} \frac{1-\rho^{c+1}}{1-\rho}$$

$$L = \frac{\rho(1-\rho)}{1-\rho^{c+1}} \frac{-(1-\rho)(c+1)\rho^c + (1-\rho^{c+1})}{(1-\rho)^2} = \frac{\rho}{1-\rho} - \frac{(c+1)\rho^{c+1}}{(1-\rho^{c+1})}$$

$$L = \frac{(\lambda/\mu)}{1-(\lambda/\mu)} - \frac{(c+1)(\lambda/\mu)^{c+1}}{\left[1-(\lambda/\mu)^{c+1}\right]}$$

Since customers who arrive when the system is full balk,

$$\lambda_a = \lambda(1 - p_c)$$

By Little's Law,

$$W = L / \lambda_a$$

Since $S(n) \sim \text{expon}(\mu)$,

$$W_S = 1 / \mu$$

Since $W = W_Q + W_S$,

$$W_Q = W - W_S$$

By Little's Law,

$$L_Q = \lambda_a W_Q$$

and

$$L_S = \lambda_a W_S$$

By the PASTA principle,

$$a_n = p_n$$

For the M/M/1/c queue, does $a_n = d_n$, $n = 0, 1, \ldots$? At steady state, customers sometimes arrive to find the system full ($a_c > 0$). However, these customers do not enter the system. As a result, no customer ever leaves the system full (d_c = 0). So, the answer is no.

8.4 The M/M/s Queue

Next we consider the M/M/s queue. Interpretation of Kendall's notation indicates the following characteristics of M/M/s queues: exponential interarrival times (rate λ), exponential service times (rate μ), s servers (but still a single queue), infinite system capacity, and infinite customer population.

We can utilize a continuous-time Markov chain to construct expressions for p_n, $n = 0, 1, \ldots$. Let $X(t)$ denote the number of customers in the system at time t.

State 0 balance equation:

$$\lambda p_0 = \mu p_1 \quad \Rightarrow \quad p_1 = \rho\, p_0 \qquad \text{where } \rho = \lambda/\mu$$

State 1 balance equation:

$$(\lambda + \mu) p_1 = \lambda p_0 + 2\mu p_2 \quad \Rightarrow \quad p_2 = \frac{\rho^2}{2} p_0$$

State 2 balance equation:

$$(\lambda + 2\mu) p_2 = \lambda p_1 + 3\mu p_3 \quad \Rightarrow \quad p_3 = \frac{\rho^3}{6} p_0$$

Continuing this process through state $s - 1$ yields

$$p_n = \frac{\rho^n}{n!} p_0$$

$n = 0, 1, \ldots , s$. State s balance equation:

$$(\lambda + s\mu) p_s = \lambda p_{s-1} + s\mu p_{s+1} \quad \Rightarrow \quad p_{s+1} = \frac{\rho^{s+1}}{s \cdot s!} p_0$$

State $s + 1$ balance equation:

$$(\lambda + s\mu) p_{s+1} = \lambda p_s + s\mu p_{s+2} \quad \Rightarrow \quad p_{s+2} = \frac{\rho^{s+2}}{s^2 \cdot s!} p_0$$

Continuing with additional states yields

$$p_n = \frac{\rho^n}{s^{n-s}\cdot s!}\,p_0$$

$n = s, s+1, \ldots$

We then take advantage of the fact that the limiting probabilities must sum to 1.

$$\sum_{n=0}^{\infty} p_n = p_0\left[\sum_{n=0}^{s-1}\frac{\rho^n}{n!}+\frac{\rho^s}{s!}\sum_{n=s}^{\infty}\left(\frac{\rho}{s}\right)^{n-s}\right] = p_0\left[\sum_{n=0}^{s-1}\frac{\rho^n}{n!}+\frac{\rho^s}{s!}\sum_{j=0}^{\infty}\left(\frac{\rho}{s}\right)^{j}\right]$$

$$\sum_{n=0}^{\infty} p_n = p_0\left[\sum_{n=0}^{s-1}\frac{\rho^n}{n!}+\frac{\rho^s}{s!}\frac{1}{1-(\rho/s)}\right] = 1 \text{ if } \frac{\lambda}{s\mu} < 1$$

$$p_0 = \left[\sum_{n=0}^{s-1}\frac{(\lambda/\mu)^n}{n!}+\frac{(\lambda/\mu)^s}{s!}\frac{1}{1-(\lambda/s\mu)}\right]^{-1}$$

$$p_n = \begin{cases} \dfrac{(\lambda/\mu)^n}{n!}\,p_0 & n = 1, 2, \ldots, s \\[2ex] \dfrac{(\lambda/\mu)^n}{s^{n-s}\cdot s!}\,p_0 & n = s, s+1, \ldots \end{cases}$$

Note that this result required that $\rho^* = \lambda/(s\mu) < 1$. Again, this condition ensures that the servers can keep up with customer arrivals. This result also allows the derivation of the remaining measures of long-run system performance.

By the PASTA principle,

$$a_n = p_n$$

Since there is no finite system capacity and customers arrive and are serviced individually,

$$d_n = a_n = p_n$$

We next construct L_Q.

$$L_Q = \sum_{n=s}^{\infty}(n-s)p_n = \sum_{n=s}^{\infty}(n-s)\frac{(\lambda/\mu)^n}{s^{n-s}\cdot s!}p_0 = \frac{(\lambda/\mu)^s}{s!}p_0\sum_{n=s}^{\infty}(n-s)(\lambda/s\mu)^{n-s}$$

$$L_Q = \frac{(\lambda/\mu)^s}{s!}p_0\sum_{j=0}^{\infty}j(\rho*)^j = \frac{(\lambda/\mu)^s}{s!}p_0\rho*\sum_{j=0}^{\infty}j(\rho*)^{j-1}$$

$$L_Q = \frac{(\lambda/\mu)^s}{s!}p_0\rho*\sum_{j=0}^{\infty}\frac{d}{d\rho*}(\rho*)^j = \frac{(\lambda/\mu)^s}{s!}p_0\rho*\frac{d}{d\rho*}\sum_{j=0}^{\infty}(\rho*)^j$$

$$L_Q = \frac{(\lambda/\mu)^s}{s!}p_0\rho*\frac{d}{d\rho*}\frac{1}{1-\rho*} = \frac{(\lambda/\mu)^s}{s!}\frac{p_0\rho*}{(1-\rho*)^2}$$

$$L_Q = \frac{(\lambda/\mu)^s}{s!}\frac{(\lambda/s\mu)}{[1-(\lambda/s\mu)]^2}p_0$$

Finally,

$$W_Q = L_Q/\lambda$$

$$W_S = 1/\mu$$

$$W = W_Q + W_S$$

$$L = \lambda W$$

and

$$L_S = \lambda W_S$$

Another key result regarding the M/M/s queue, **Burke's theorem**, addresses the departure of customers.

> At steady state, departures from the M/M/s queue occur according to a Poisson process having rate λ.

Since the M/M/1 queue is a special case of the M/M/s queue, Burke's theorem also applies to the M/M/1 queue.

8.5 The M/M/s/c Queue

Next we consider the M/M/s/c queue. Interpretation of Kendall's notation indicates the following characteristics of M/M/s/c queues: exponential interarrival times (rate λ), exponential service times (rate μ), s servers, system capacity of c, and infinite customer population. We assume that $c > s$.

We can utilize a continuous-time Markov chain to construct expressions for p_n, $n = 0, 1, \ldots, c$. Let $X(t)$ denote the number of customers in the system at time t. Note that $X(t) \in \{0, 1, \ldots, c\}$ for all $t \geq 0$.

Note that states $0, 1, \ldots, c-1$ have identical balance equations to the M/M/s queue. Therefore,

$$p_n = \frac{\rho^n}{n!} p_0$$

$n = 0, 1, \ldots, s$, and

$$p_n = \frac{\rho^n}{s^{n-s} \cdot s!} p_0$$

$n = s, s+1, \ldots, c$.

We then take advantage of the fact that the limiting probabilities must sum to 1.

$$\sum_{n=0}^{c} p_n = p_0 \left[\sum_{n=0}^{s} \frac{(\lambda/\mu)^n}{n!} + \frac{(\lambda/\mu)^s}{s!} \sum_{n=s+1}^{c} \left[\frac{(\lambda/\mu)}{s} \right]^{n-s} \right] = 1$$

$$p_0 = \left[\sum_{n=0}^{s} \frac{(\lambda/\mu)^n}{n!} + \frac{(\lambda/\mu)^s}{s!} \sum_{n=s+1}^{c} \left[\frac{(\lambda/\mu)}{s} \right]^{n-s} \right]^{-1}$$

Next, we derive L_Q as we did with the M/M/s queue.

$$L_Q = \sum_{n=s}^{c}(n-s)p_n = \sum_{n=s}^{c}(n-s)\frac{\rho^n}{s^{n-s}s!}p_0 = \frac{\rho^s p_0}{s!}\sum_{j=0}^{c-s} j(\rho*)^j = \frac{\rho^s \rho* p_0}{s!}\sum_{j=0}^{c-s} j(\rho*)^{j-1}$$

$$L_Q = \frac{\rho^s \rho* p_0}{s!}\sum_{j=0}^{c-s}\frac{d}{d\rho*}(\rho*)^j = \frac{\rho^s \rho* p_0}{s!}\frac{d}{d\rho*}\sum_{j=0}^{c-s}(\rho*)^j$$

$$L_Q = \frac{\rho^s \rho* p_0}{s!}\frac{d}{d\rho*}\frac{1-(\rho*)^{c-s+1}}{1-\rho*}$$

$$L_Q = \frac{\rho^s \rho* p_0}{s!}\frac{-(1-\rho*)(c-s+1)(\rho*)^{c-s} + \left[1-(\rho*)^{c-s+1}\right]}{(1-\rho*)^2}$$

$$L_Q = \frac{\rho^s \rho* p_0}{s!(1-\rho*)^2}\left[1-(\rho*)^{c-s} - (c-s)(\rho*)^{c-s}(1-\rho*)\right]$$

The remaining performance measures follow in the typical manner.

$$\lambda_a = \lambda(1-p_c)$$

$$W_Q = L_Q/\lambda_a$$

$$W_S = 1/\mu$$

$$W = W_Q + W_S$$

$$L_S = \lambda_a W_S$$

$$L = \lambda_a W$$

8.6 The M/G/1 Queue

Next we consider the M/G/1 queue. Interpretation of Kendall's notation indicates the following characteristics of M/G/1 queues: exponential interarrival times (rate λ), general service times (no assumed probability distribution on service time), one server, infinite system capacity, and infinite customer population.

Since service times are not exponentially distributed, we cannot use continuous-time Markov chains to analyze the behavior of the M/G/1 queue. However, there are well-known results for the long-run performance of the M/G/1 queue assuming the mean and variance of service time are known.

For the M/G/1 queue,

$$W_Q = \frac{\lambda(\text{second moment of service time})}{2(1-\lambda W_S)}$$

$$L_Q = \lambda W_Q$$

$$W = W_Q + W_S$$

and

$$L = \lambda W$$

Note that all of these values increase as the variance of the service time distribution increases. From a customer's perspective, increases in these measures are undesirable.

For any single-server queueing system, the server alternates between busy and idle periods. For the M/G/1 queue, an idle period ends when a customer arrives. So, the length of an idle period is an exponential random variable having rate λ. Analysis of busy periods is more difficult. However, there are well-known results regarding busy periods of the M/G/1 queue.

Consider the M/G/1 queue. If B denotes the long-run average length of a busy period and C denotes the long-run average number of customers served during a busy period, then

$$B = \frac{W_S}{1-\lambda W_S}$$

and

$$C = \frac{1}{1-\lambda W_S}$$

There are also well-known results regarding the long-term performance of the M/G/1 queue when customers arrive in batches.

> Consider the M/G/1 queue, except that each arrival consists of N customers where N is a random variable (mean and variance known). First, note that $\lambda_a = \lambda\, E(N)$. In addition,

$$W_Q = \frac{\dfrac{W_S}{2E(N)}\left[E\left(N^2\right)-E\left(N\right)\right]+\dfrac{\lambda}{2}E\left(N\right)\left(\text{second moment of service time}\right)}{1-\lambda E\left(N\right)W_S}$$

$$W = W_Q + W_S$$

$$L_Q = \lambda_a W_Q$$

and

$$L = \lambda_a W$$

8.7 Networks of Queues

In many industrial scenarios, queues are connected in networks. The simplest example of a queueing network is a serial production (assembly) line. In an assembly line, parts wait for a machine, get processed by that machine, and then proceed to the queue for the next machine. We consider some very basic concepts regarding queueing networks known as **Jackson networks**. We focus on **open systems**, where new customers are permitted to enter the system and existing customers are permitted to depart the system. Many advanced texts include discussion of **closed systems**, where new customers cannot enter and existing customers never leave.

Consider the following two-server queueing system.

> Customers arrive at the system according to a Poisson process having rate λ. Upon arrival, they enter the queue for server 1 (exponential service with rate μ_1). After completing service by server 1, the customer proceeds to the queue for server 2 (exponential service with rate μ_2). After completing service by server 2, the customer departs the system.

If we consider only server 1, then the appropriate queueing model is an M/M/1 queue. According to Burke's theorem, customers depart server 1 according to a Poisson process having rate λ. Therefore, if we consider only server 2, then the appropriate queueing model is an M/M/1 queue. We can supplement these conclusions with four new results.

New Result 1: The number of customers in "subsystem 1" is independent of the number of customers in "subsystem 2."

New Result 2: Let (n, m) denote the state of the system where n denotes the number of customers in subsystem 1 and m denotes the number of customers in subsystem 2. If $p_{n,m}$ denotes the steady-state probability of being in state (n, m), then

$$p_{n,m} = \left(\frac{\lambda}{\mu_1}\right)^n \left(1 - \frac{\lambda}{\mu_1}\right)\left(\frac{\lambda}{\mu_2}\right)^m \left(1 - \frac{\lambda}{\mu_2}\right)$$

$n = 0, 1, \ldots, m = 0, 1, \ldots$.

New Result 3:

$$L = \frac{\lambda}{\mu_1 - \lambda} + \frac{\lambda}{\mu_2 - \lambda}$$

New Result 4:

$$W = \frac{1}{\mu_1 - \lambda} + \frac{1}{\mu_2 - \lambda}$$

The preceding example can be extended to a system of any number of queues. And, it can be generalized.

Consider a system of k servers. New customers arrive at server i, $i = 1, 2, \ldots, k$, according to a Poisson process having rate r_i. Let μ_i denote the exponential service rate at server i, $i = 1, 2, \ldots, k$. Once a customer finishes service by server i, $i = 1, 2, \ldots, k$, he joins the queue for server j, $j = 1, 2, \ldots, k$, with probability P_{ij}.

We can identify four new results for this general system corresponding to the four from the two-server system.

New Result 1: If we let λ_j denote the total arrival rate of customers at the queue for server j, $j = 1, 2, \ldots, k$, then $\lambda_1, \lambda_2, \ldots, \lambda_k$ can be obtained as the solution of

$$\lambda_j = r_j + \sum_{i=1}^{k} \lambda_i P_{ij}$$

$j = 1, 2, \dots, k$

New Result 2: If L_j denotes the number of customers in subsystem j, then $L_1, L_2, \dots, L_k$ are independent random variables and, in the long run,

$$\Pr(L_1 = n_1) = \left(\frac{\lambda_j}{\mu_j}\right)^{n_j}\left(1 - \frac{\lambda_j}{\mu_j}\right)$$

$j = 1, 2, \dots, k$, where $n_1, n_2, \dots, n_k$ are all non-negative integers. Therefore,

$$\Pr(L_1 = n_1, L_2 = n_2, \dots, L_k = n_k) = \prod_{j=1}^{k}\left(\frac{\lambda_j}{\mu_j}\right)^{n_j}\left(1 - \frac{\lambda_j}{\mu_j}\right)$$

New Result 3:

$$L = \sum_{j=1}^{k} \frac{\lambda_j}{\mu_j - \lambda_j}$$

New Result 4:

$$W = \frac{\sum_{j=1}^{k} \frac{\lambda_j}{\mu_j - \lambda_j}}{\sum_{j=1}^{k} r_j}$$

Homework Problems

8.1 Queueing Basics

(1) Consider a D/D/1 queue such that $T(n) = 2$ minutes and $S(n) = 1$ minute, $n = 1, 2, \dots$. Evaluate the nine fundamental measures of long-run queueing system performance.

8.2 The M/M/1 Queue

(1) Consider an M/M/1 queue having $\lambda = 3$ arrivals per minute and $\mu = 4$ customers per minute. Compute the following steady-state quantities.

(a) the proportion of time during which three customers are in the system

(b) the average number of customers in the system

(c) the average amount of time a customer spends in the system

(d) the average amount of time a customer spends waiting in queue

(e) the average number of customers waiting in queue

(f) the average number of customers in service

(g) the proportion of arriving customers who find three customers in the system

(h) the proportion of departing customers who leave three customers in the system

(i) the proportion of time that the server is busy

(2) Determine the probability distribution of the time a customer spends in an M/M/1 queue.

(3) Consider an M/M/1 queue, and suppose that a customer just departed. What is the expected value of Y, the time until the next departure?

(4) For an M/M/1 queue, determine the CDF of the amount of time a customer spends waiting in queue.

8.3 The M/M/1/*c* Queue

(1) Consider an M/M/1/3 queue having $\lambda = 3$ arrivals per minute and $\mu = 4$ customers per minute. Compute the following steady-state quantities.

(a) the proportion of time during which three customers are in the system

(b) the average number of customers in the system

(c) the average amount of time a customer spends in the system

(d) the average amount of time a customer spends waiting in queue

(e) the average number of customers waiting in queue

(f) the average number of customers in service

8.4 The M/M/s Queue

(1) Consider an M/M/3 queue having arrival rate $\lambda = 10$ customers per minute and service rate $\mu = 4$ customers per minute. Compute the following steady-state quantities.

(a) the probability that the system is empty

(b) the probability that an arriving customer finds two customers in the system

(c) the probability that a departing customer leaves six customers in the system

(d) the average number of customers in the queue

(e) the average amount of time a customer spends in the system

(f) the average number of customers in the system

(g) the average number of idle servers

(2) Prove Burke's theorem for the M/M/1 queue.

8.5 The M/M/s/c Queue

(1) Consider an M/M/2/5 queue having $\lambda = 6$ customers per hour and $\mu = 9$ customers per hour. Compute the following steady-state quantities.

(a) the proportion of time during which there are five customers in the system

(b) the average number of customers waiting for service

(c) the average number of busy servers

(d) the average number of customers in the system

(e) the average amount of time a customer spends waiting for service

(f) the average amount of time a customer spends in the system

(g) the average number of customers lost per hour

(h) the average amount of time a customer spends in the system (including customers who balk)

8.6 The M/G/1 Queue

(1) For an M/M/1 queue, what are the steady-state values of:

(a) the average length of a busy period?

(b) the average number of customers processed during a busy period?

(2) Customers arrive at a single-server queue according to a Poisson process having $\lambda = 9$ customers per hour. The time required to process a single customer is a 2-Erlang random variable having a rate of 0.4 customer per minute. Compute the following steady-state quantities.

(a) the average amount of time a customer spends in the system

(b) the average number of customers in the system

(c) the average length of a busy period

(3) Customers arrive at a single-server queue in groups. Group arrivals occur according to a Poisson process having $\lambda = 3.5$ groups per hour. The number of customers in a group is a geometric random variable having a mean of 2. The server processes customers sequentially. The average time required to process a single customer has a mean of 8 minutes and a standard deviation of 2 minutes. Compute the following steady-state quantities.

(a) the average amount of time a customer spends waiting in queue

(b) the average number of customers waiting in queue

(c) the average number of customers in the system

8.7 Networks of Queues

(1) Consider a three-server system. Customers arrive at the system according to a Poisson process having $\lambda = 2.5$ customers per minute. Upon arrival, customers enter the queue for server 1 (exponential service rate of 3 customers per minute). After completing service by server 1, the customer proceeds to the queue for server 2 (exponential service rate of 3.5 customers per minute). After completing service by server 2, the customer proceeds to the queue for server 3 (exponential service rate of 2.8 customers per minute). After completing service by server 3, the customer departs the system. Assume all three queues have infinite capacity. Compute the following steady-state quantities.

(a) the probability that two customers are in the system

(b) the average number of customers in the system

(c) the average amount of time a customer spends waiting in queue

(2) A manufacturing system that contains three machines produces three types of parts upon order. Customer order arrive according to a Poisson process having a rate of 0.15 orders per hour. Of these orders, 50% are for part 1, 25% are for part 2, and 25% are for part 3. Part 1 requires processing time on machines 1, 2, and 3 (in that order). Part 2 requires processing time only on machines 2 and 3 (in that order), and part 3 requires processing time on machines 3 and 1 (in that order). The time required to process a part on machine i is an exponential random variable having a rate of μ_i parts per hour. Note that $\mu_1 = 0.25$ part per hour, $\mu_2 = 0.35$ part per hour, and $\mu_3 = 0.4$ part per hour. Compute the following steady-state quantities.

(a) the average amount of time a part spends in the system

(b) the average number of parts in the system

Bibliography

Bennett, D. J. 1999. *Randomness*. Cambridge, MA: Harvard University Press.

Birnbaum, Z.W. On the importance of different components in a multi-component system in Multivariate Analysis, P.R. Krishnaiah, Ed:, Academic Press, 1969, vol. 11.

Burke, P. J. 1956. The Output of a Queueing System. *Operations Research* 4 (6): 699–704.

Kendall, D. G. 1953. Stochastic Processes Occurring in the Theory of Queues and their Analysis by the Method of the Imbedded Markov Chain. *Annals of Mathematical Statistics* 24 (3): 338–354.

Little, J.D.C. A proof of the queuing formula $L = \lambda W$. Operations Research, vol. 9, 1961 pp. 383–387.

Ross, S. M. 2000. *Introduction to Probability Models*, 7th ed. San Diego, CA: Harcourt Academic Press.

Wald, A. 1944. On Cumulative Sums of Random Variables. *Annals of Mathematical Statistics* 15 (3): 283–296.

Index

D

E

F

G

H

I

J

K

L

N

O

P

Q

R

S